The Design and Installation Guide for Roof Tiling

The Design and Installation Guide for Roof Tiling is the definitive guide to obtaining successful results in pitched roofing construction using clay or concrete roof tiles. It was written by a team of experts from the Roof Tile Association (RTA) and the National Federation of Roofing Contractors (NFRC), the representative trade associations for pitched roofing. This book is in line with British and European Standards, the latest best site practices and industry recommendations to ensure readers are receiving the most up-to-date and accurate information available in the field.

Based on actual teaching practice, this book is highly illustrated throughout to increase the accessibility of the text for the reader in their exploration of the practical aspects of roofing. It also includes an extensive glossary of roofing terms for ease of reference. It brings together a comprehensive collection of the design, material specifications and workmanship requirements to construct a successful tiled pitched roof, including:

- up-to-date design, product and workmanship standards
- current best site practice
- advances in health and safety
- current typical scope of works for a roofing contractor
- broad knowledge of the overall building envelope performance; specifically, the increasing insulation and airtightness requirements.

Students following diploma (Foundation, Intermediate and Advanced levels) and apprenticeship routes will find this book to be an invaluable reference source of information that will accompany them throughout their studies. Building professionals concerned with the design, detailing and specification of roofing will also find this book to be an essential reference.

Note: The guidance contained in this book is about roof-tile products and accessories to the Great Britain (GB) market. The GB market refers to England, Wales and Scotland. It is a summary of the provisions of the relevant legislation and guidance, but it is not comprehensive. You should rely on the most current provisions of the appropriate country of legislation in preference to what is summarised in this guidance.

Recommendations and guidance given is designed to be informative. Conformity should be checked, and the product manufacturer's guidance and recommendations should be sought for the specific use or application.

All legislation or regulation, including British and European Standards referred to, were current at the time of publication.

The Roof Tile Association (RTA) represent the UK's leading manufacturers of pitched roof products and solutions. They are the central representative body establishing a common viewpoint for clay and concrete roof-tile manufacturers.

Their combined knowledge of traditional pitched tiled roofs is now available to share with anyone looking to create the most environmentally responsible, thermally efficient and architecturally inspiring roofs in the UK.

The National Federation of Roofing Contractors (NFRC) is the largest and most influential roofing trade association in the UK, promoting quality contractors and products and ensuring that its members are at the forefront of all roofing developments.

NFRC actively ensures that all members offer high standards of workmanship and sound business practice through a strict code of practice and vetting procedure, including site inspections. They also offer training and technical advice and represent member interests to the wider construction industry and government.

The Design and Installation Guide for Roof Tiling

RTA and NFRC

Routledge
Taylor & Francis Group
LONDON AND NEW YORK

Roof Tile Association

NFRC
LEADING ROOFING EXCELLENCE

Designed cover image: Kindly supplied by RTA members

First published 2025
by Routledge
4 Park Square, Milton Park, Abingdon, Oxon OX14 4RN

and by Routledge
605 Third Avenue, New York, NY 10158

Routledge is an imprint of the Taylor & Francis Group, an informa business

British Library Cataloguing-in-Publication Data
A catalogue record for this book is available from the British Library

ISBN: 978-1-032-05314-1 (hbk)
ISBN: 978-1-032-05313-4 (pbk)
ISBN: 978-1-003-19699-0 (ebk)

DOI: 10.1201/9781003196990

Typeset in Helvetica Neue
by Apex CoVantage, LLC

Contents

Abbreviations

ABIS	As Built In Service
ADT	As Designed Theoretical
AVCL	Air & Vapour Control Layer (membrane)
BBA	British Board of Agrément
BRE	Building Research Establishment
BS	British Standards (BS EN are the British standard implementation of European standards)
CAD	Computer Aided Design
CAWS	Common Arrangement of Work Sections
CITB	Construction Industry Training Board
COSHH	Control of Substances Hazardous to Health
CPET	Central Point of Expertise on Timber
EN	European Standards
FSC	Forest Stewardship Council
GRP	Glass reinforced plastic or glass reinforced polymer
HR	High (water vapour) resistance
HSE	Health and Safety Executive
ISO	International Standards Organisation
LABC	Local Authority Building Control
LR	Low (water vapour) resistance or Breathable membrane
LSA	Lead Sheet Association
LSTA	Lead Sheet Training Academy
MCWP	Mast climbing work platform
MEWP	Mobile elevated work platform
MNs/g	Mega-Newton seconds per gram (vapour resistance of a material)
NASC	National Access and Scaffolding Confederation
NBCRM	Non-bitumen coated roofing membranes
NFRC	National Federation of Roofing Contractors
NHBC	National House Building Council
PEFC	Programme for the Endorsement of Forest Certification
PV	Photovoltaic roofs (solar panels on roofs)
PDF	Portable Document Format files containing images and text that can be digitally viewed or printed.
RBM	Reinforced bitumen membranes

RIBA	Royal Institute of British Architects
RoofCERT	The NFRC roofing accreditation scheme for roofers
RTA	Roof Tile Association
SIPS	Structural insulated panels
SVP	Soil vent pipe
UKAS	United Kingdom Accreditation Service
UV	Ultraviolet (radiation from sunlight)
UVA	Most common UV-ray comprising long (damaging) wavelength
UVB	Less common UV-ray comprising short (less damaging) wavelength
VCL	Vapour control layer
VPM	Vapour permeable membranes
VPU	Vapour permeable underlays.

Diagrams

All diagrams shown in this book are for illustration purposes and are typical representations of detail. Unless otherwise shown, original copyright is owned, and permission to use granted, by the Roof Tile Association (including The Clay Roof Tile Council and Surevent), Marley Limited, Wienerberger Limited (Sandtoft), Russell Building Products Limited (Russell Roof Tiles), Hinton Perry & Davenhill Ltd (Dreadnought Tiles), Tudor Roof Tile Co. Limited.

Acknowledgements

Thank you to those RTA members and the NFRC team who formed the working group and have contributed to this book. Their tremendous effort and expertise have resulted in this thorough and informative guide.

A special thank you to the 2021–24 Chair of the RTA, Paul John Lythgoe, who persevered in collating all the information together and bringing it to a printable format which you now read.

CHAPTER 1

Foreword

Pitched roofing is elegantly simple and effective, which is why it has survived for several thousand years. Although technically having evolved considerably, the principles are still fundamentally the same: most importantly, to be appropriate for the building location and to ensure stability and security.

Pitched roofs have been part of the built environment for around 5,000 years demonstrating what a sound and effective construction method pitched roofing is. Whilst the earliest forms of pitched roofs were laid in simple materials such as leaf, bamboo, turf and stone, the last couple of centuries have seen a shift toward alternative materials with roof tiles formed of clay and concrete.

Today, modern manufacturing processes have resulted in the ability to produce precision-engineered roof tiles, which continue to offer versatile, cost-effective and visually appealing roofing solutions.

Although tiles are available in a wide range of materials such as wood, slate, stone and synthetic rubber, the colour, performance and ease of installation have established clay and concrete roof tiles as the most preferred choice of roof covering around the world.

1.1 What is a pitched roof

A pitched roof is the topmost covering of a building with slopes greater than 10°, typically two sloping sides meeting at a top ridge. These double-pitched roofs are the most common roof built in the UK, however, many other types ranging from simple single slope to complex multiple slopes are commonplace.

The pitch is the angle of the roof rise over its horizontal run (usually rafters vertical rise divided by horizontal span) and is often also referred to as the slope of the roof.

1.2 Versatility and shape

The simple principles of pitched roofing do not limit artistic expression, and a huge range of roof shapes and types can be accomplished with a pitched roof.

1.3 Structural integrity

Pitched roofs are very effective at carrying and distributing not only the weight of the chosen covering, but also the additional loads applied to it by environmental factors such as wind, rainwater and snow. Gravitational forces naturally clear precipitation off a pitched roof.

DOI: 10.1201/9781003196990-1

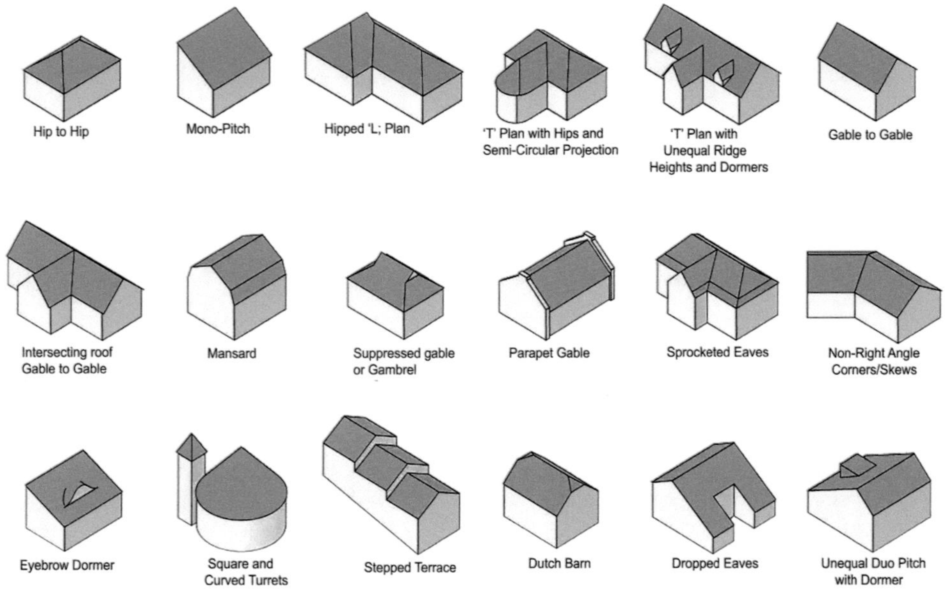

Figure 1.1 Examples of pitched roof types

Common pitched roof coverings such as clay and concrete roof tiles compliment a pitched roof's structural performance and provide excellent resistance to challenging environmental conditions. However, both the roof substructure and its tile covering need to be installed correctly following building regulations and codes of practice in order to preserve the roof's integrity and reduce the risk of structural failure.

1.4 A sustainable platform

The pitched roof system is an ideal platform for sustainable technologies such as photovoltaic (PV) or other solar thermal systems, and this should be considered by architects when designing roof orientation and pitch. A pitched roof can be used effectively to promote the collection of rainwater; in many cases, it can support a rainwater harvesting system. A pitched roof enables the owner to reuse substantially more rainwater than a flat roof.

1.5 Rainwater drainage

The pitched roof is exceptionally good at directing water from the roof surface into the guttering and away from the building. Maintenance checks and gutter cleaning are needed since any blockages in runoff can lead to costly water damage.

1.6 Indoor comfort

A pitched roof protects a house's facade from solar radiation in summer, thus avoiding overheating; in winter it enables the low sun to heat the inside and protects against wind-driven rain. A pitched roof significantly improves the thermal efficiency of the building thanks to the natural ventilation present underneath the top roof layer, which acts as a buffer zone to the external environment. This ensures maximum home comfort both in winter and summer.

The ventilated cavity present in a pitched roof helps to evacuate moisture from the roof and prevents internal condensation from forming. A well-designed and executed pitched roof provides excellent acoustic insulation thanks to its combination of sound-absorbing materials.

1.7 Extra space for less

One of the most attractive features of choosing a pitched roof is the extra space it provides at a relatively low cost. The space created by a pitched roof can be used for a range of purposes, most commonly as additional storage.

There is a growing trend toward people choosing to extend their property, rather than move. Many homeowners view the space underneath a pitched roof as an opportunity to create additional living space affordably. In most cases, the creation of additional living space, often referred to as a 'loft conversion', offers a cheaper alternative to an external extension because the space already exists, and therefore the cost lies in converting rather than building from scratch.

1.8 Long lifespan, minimum maintenance

If constructed properly, a well-designed pitched roof offers a long life span, with materials that are very durable and weather-resistant. The internal accommodation will be efficiently insulated and will not suffer from extremes of temperature.

For the UK climate, the structural design of a pitched roof brings with it many practical advantages that will help to reduce the potential for damage and costs. Insulation in a pitched roof is not vulnerable to damage caused by water, moisture or vegetation; this helps ensure the long-term performance of the building.

There are significant advantages to specifying a pitched roof building design; it delivers a range of social, economic and environmental benefits to both the homeowner and the developer.

1.9 Training as a roofer

Successful roofing requires a mixture of quality products and quality installation. Skilled craftsmanship is needed when laying a clay or concrete tile roof, and the roofing industry recommends training and certification.

The Construction Industry Training Board (CITB), NFRC and other stakeholders develop and provide educational standards, certification, qualifications and training courses for roofing contractors and their operatives which comprise the RoofCERT scheme, National/Scottish Vocational Qualification (NVQ/SVQ) and apprenticeships.

1.10 Glossary of terms

This section explains some of the most familiar terms used in the clay roof tile market.

Key terms – roof

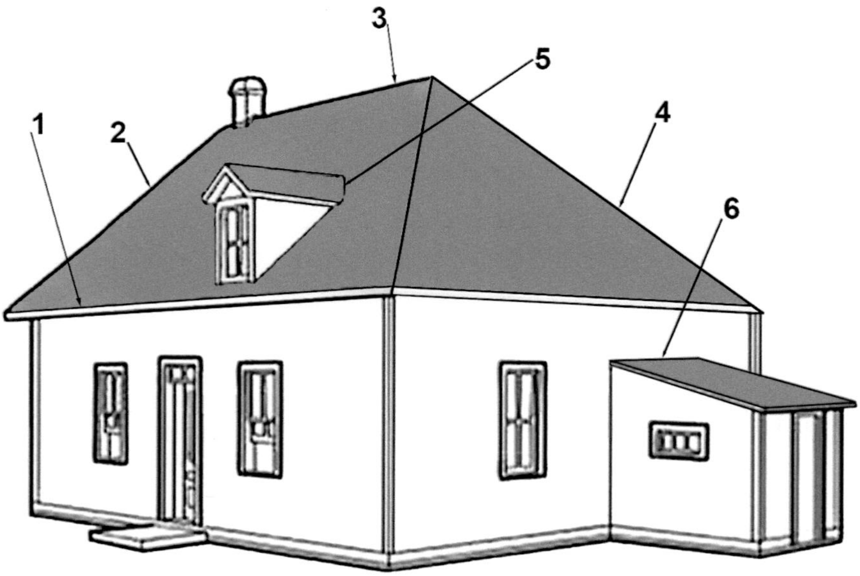

Figure 1.2 Key roof terms

1. Eaves

The eaves are where the roof drains into the gutter and is the point where there is most water, therefore careful attention to detail is important. If there is a gap below the tiles at the eaves of 16 mm or more then eaves fillers are required for profile tiles to prevent access by birds and rodents. The eaves play a major part in ventilating the roof space.

2. Verges

Where the roof starts and finishes at a gable wall this is commonly referred to as the roof verge. Traditionally verges are finished with a mortar bedding but to avoid the need for future maintenance tiles can be dry fixed using a cloaked verge tile or dry verge system.

3. Ridge

The ridgeline finishes of the roof at the top and the ridge tiles can be traditionally mortar bedded or dry fixed using a dry ridge system. Ridges also offer great design opportunities using decorative ridges and finials. As with the eaves, the ridge can play a major part in ventilating the roof space.

4. Hip

A hip is where two roof slopes meet, forming a junction from which the water runs away. Like the ridge, the hip junction can either be mortar bedded or finished dry using a dry hip system.

5. Valley

A valley is where two roof slopes meet, forming a junction into which the water runs. Valleys are made watertight using GRP (glass reinforced plastic) or lead to line the valley trough. With plain tiles, purpose-made valley tiles can be used.

6. Abutment

Where a roof meets a wall or other vertical projection at the top of the tiling, this is referred to as a **top** abutment. Top abutments are finished using a lead cover flashing.

A **side** abutment is where a roof verge meets a wall that rises above the tiling. Profiled tiles can be weathered using a cover flashing. For flat tiles, a secret gutter should be used.

Key terms – tiles

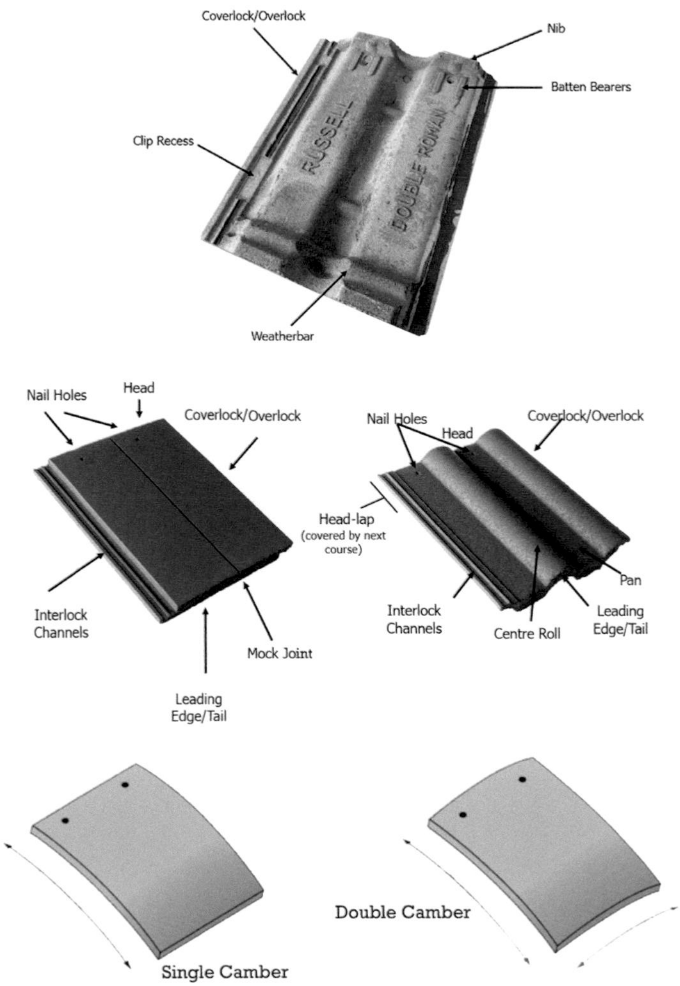

Figure 1.3 Key tile terms

Terminology

ABUTMENT The junction of a roof surface with a wall, or any other structural feature which arises above it.

AIR & VAPOUR CONTROL LAYER Membrane that prevents water vapour from warm air passing through it.

APEX The highest point of the roof where the two slopes meet.

ARRIS HIP A hip tile ordered specifically to suit the roof pitch.

BABY RIDGE A small version of the half-round ridge that is suitable for use on hips and bays.

BACK Underside of tile when laid on a roof.

BACK BEDDING Mortar applied under ridge or hip tiles, which is not visible.

BARGE BOARD A board fixed along the edge of a gable.

BATTENS Horizontal, small-section strips of timber graded to BS 5534 on which tiles or slates are laid.

BATTEN GAUGE Often referred to as 'the gauge'. This is the measurement determined by the tile for batten spacing. It is the distance from the top of one batten to the top of the next batten below. Correct batten spacing is essential for a weather-tight roof.

BATTEN VOID The area above the underlay and below the roof covering.

BLOCK END RIDGE A special ridge tile for use at the gable end. Often used with cloaked verge tiles.

BONDING GUTTER A preformed length of flashing designed to weather the joint between two roofs with different types of roof coverings.

BONNET HIP Rounded hip tile used in plain tiling.

BROKEN/HALF BOND A way of laying tiles so that the edge of each tile is above the middle of the tile in the course below (¼ and ¾ bond for mock joint tiles).

CLOAK VERGE TILE A special tile fitting with a section which turns down at the verge.

CLIP An aluminium, steel or plastic device to secure tile to the roof structure beneath.

COLD ROOF A roof designed with the insulation at ceiling level.

COUNTER BATTENS Timbers fixed vertically between the battens and the surface below. A batten mounted vertically up the roof along the lines of the rafters. These are normally used where the roof frame has been boarded to give a space when the underfelt and battens are fixed.

COURSE A horizontal row of tiles or slates.

CONDENSATION Where hot air meets cold air, condensation is formed. For example, where heat rising from the inside of the house meets the cold underside of the roof covering, condensation occurs.

DECORATIVE RIDGE A ridge tile available in a variety of shaped crests.

DENTIL SLIPS Small rectangular pieces of plain tile bedded into the pans of deep-profile tiles to reduce the amount of mortar required for bedding hip and ridge tiles.

DORMER Roof window that projects out vertically from the roof slope.

DOUBLE-LAP Description for roofing materials without interlocking channels, where each course overlaps the course next but one below (see also single-lap).

DOUBLE PANTILE Single-lap tile moulded to two pans in section.

DOUBLE CAMBER A tile arched both horizontally and vertically to break up the dominance of the course line and give the roof an undulating look.

DRY RIDGE A roof ridge which is mechanically fixed without mortar.

DRY VERGE A roof verge which is mechanically fixed without mortar.

DRY VALLEY A preformed valley liner which is mechanically fixed without mortar.

DRY HIP A system which allows for a hip to be mechanically fixed without mortar.

EAVES The lower/draining edge of a roof.

EAVES/TOPS TILE Shorter tiles used with plain tiles in a single course under the standard tile to give a double course of tiling at the eaves. A single course of short tiles is also used on a top course on both sides of the ridge.

EAVES FILLER A component that fills the space under the roll of a roof tile at the eaves to keep out birds, vermin and insects.

EFFLORESCENCE The formation of a white crystalline deposit on the surface of a tile, caused by mineral migration to the surface.

EYEBROW Tiles swept up from either side over a dormer window.

FASCIA BOARD The board attached vertically to the rafter ends at the eaves, the wall plate or the wall face.

FELT/MEMBRANE (Also known as underlay or sarking felt). A water-resistant barrier supplied in rolls and laid over rafters or counter battens.

FINIAL A decorative fitting used at the end of a ridge or at the highest point of a roof.

FIXED-GAUGE TILES Tiles where the head-lap is either fixed by mitred corners, or restricted by anti-capillary bars at the head and tail, often requiring a cut short course at the ridge.

FIXINGS Nails, pegs, hooks and screws used to fix tiles and battens.

FLASHING A sheet of metal, usually lead or aluminium, which protects a joint from water penetration.

GABLE The vertical triangular section of wall above the level of the eaves and below the sides of a pitched roof.

GABLE END The end wall where two verges meet.

GAUGE The distance between the top of one batten and the top of the next, equal to the length of tile exposed (margin) after it has been installed.

GUTTER (BACK) A gutter formed at the back of a chimney, or any other structure which penetrates the roof.

GUTTER (SECRET) A gutter formed at an abutment and effectively hidden from sight. (As opposed to Side Gutter, which is exposed to view.)

GUTTER (VALLEY) A visible gutter running down the valley.

GRANULAR/SANDFACED A surface application of sand or small chippings.

HANDMADE A plain tile made by hand for additional aesthetic requirements.

HANGING TILES A general term which is applied to tiles fixed to vertical walls.

HANGING LENGTH The distance measured from the underside of the hanging nibs to the bottom edge of a tile, used to aid setting out at the eaves to maintain correct overhang distance in relation to the gutters.

HALF BOND Also known as broken or cross bond. Describes the laying pattern where the tile/slates are laid halfway across the course below (all double-lap materials).

HEAD Top edge of the tile as laid.

HEAD-LAP The measurement of the overlap of one course of slates or tiles over the course below (used to work out the batten gauge). Usually expressed as a maximum or minimum measurement in millimetres, defined by the tile or slate manufacturer.

HEADLOCK Tile transverse interlocking system, consisting of ribs and grooves to improve the resistance to water ingress.

HIP The meeting of two pitched roof surfaces which meet at an external angle.

HIP BOARD The board along the line of a hip, from the fascia to the ridge of the pitch.

HIP IRON A metal strap bent to form a stop for the hip covering and screwed to the lower end of a hip rafter.

HIP TILE A fitting designed to cover the hip intersection of a pitched roof of a given pitch.

HIP END A sloping end to a pitched roof which is covered with slates or tiles.

HIP END RIDGE A ridge tile that is used at the end of the ridge, where it meets two hips. Suitable for use with plain tiles with bonnet, arris or mitred hips, also suitable for use with slates with mitred hips.

HYBRID ROOF A combination of cold and warm roofs on the same roof slope.

INTERLOCKING TILE A single-lap tile designed to connect with adjoining tiles by close-fitting weather bars/rain channels.

LATH Old term for the wooden support for the tiles – nowadays, 'batten' is the term most used.

LAP The amount by which a tile overlaps the course below it – or, in the case of plain tiles, the course next but one.

LEAN-TO ROOF A single-pitched roof slope, which meets with a wall abutment along the top edge.

LEAD SLATE Provides a weather-tight joint where flue pipes and soil stacks penetrate a roof covering.

LEAD REPLACEMENT FLASHINGS Proprietary flashings designed to replace lead.

MITRE Materials cut so they form a close fit.

MANSARD ROOF A roof having two slopes on both sides and ends, the lower slopes being steeper in pitch.

MARGIN The length of tile visible once laid (equal to the batten gauge).

MATHEMATICAL TILES Hanging tiles that give the appearance of brick cladding.

MOCK JOINT Artificial line that gives a small format appearance to a tile.

MONOPITCH ROOF A pitched roof with a single slope from eaves to ridge.

NIBS Projecting lugs on the underside of a tile near the head, which locate the tile on the battens.

NATURAL WEATHERING The process of external building materials changing in appearance according to their environment.

OVERHANG Distance that the undercloak extends over the verge or the distance by which tiles extend over the fascia board into the gutter.

OVER-FASCIA VENT Provides discrete low-level ventilation.

OVERLOCK The part of interlocking tiles that goes over the edge of the adjacent tile. The underlock its opposite.

PANTILE Single-lap tile moulded to a flat S-shape in section.

PITCH The angle (in degrees) of the roof to the horizontal (normally applies to the rafters).

PLAIN TILE A small, slightly cambered roofing tile, typically size 265 mm x 165 mm, usually with nibs and nail holes.

PLANE A roof plane is the flat surface area of the roof that is at an angle having four separate edges.

PEG Wooden or metal pin used to hang tiles from battens.

PEG TILE A tile similar to a plain tile but without nibs and with holes for pegs.

PERP LINE Guide line struck up the roof to position tiles perpendicularly.

POINTING/BEDDING The use of mortar or cement to fill the gap between the tile and the undercloak at the verges.

PURLIN A horizontal length of timber that provides support to rafters.

RAFTER A vertical, sloping timber used to form the shape of the roof – the side of a truss.

RAFTERS Sloping structural timbers or steel which support the roof.

RAKE The portion of a roof that extends past the exterior wall at the gable end and spans from the eave to the ridge. Its purpose is to provide additional protection.

RAKING CUT A diagonal cut across courses or rows of tiles.

RECONSTITUTED SLATE A modern, manufactured product which consists of crushed natural slate which is bound together using resins.

RIDGE The junction of two slopes forming the apex of a pitched roof.

RIDGE TILE A fitting designed to cover the apex of a pitched roof of a given pitch.

RIDGE VENT TERMINAL A ridge tile which incorporates a ventilation outlet.

ROD Timber (nowadays, usually a batten) marked with the gauging for tile courses and used to mark the courses on the rafters.

ROOF JUNCTIONS The detail where two or more roof slopes meet, for example, a ridge line, a hip or a valley.

ROOFLIGHT Window in the plane of a roof.

SADDLE A piece of impervious flexible sheet material (usually lead) dressed to shape, fitted to provide weather protection.

SARKING FELT Another name for roofing underfelt/underlay.

SARKING BOARD Sarking is wood boarding used under tiles or slates to provide support to the underlay, commonly used in Scotland.

SETTING OUT Ensuring a consistent appearance of the courses, considering the required head-lap and dimensions of the roof (see section 6).

SIDE-LAP The distance which one tile/slate is offset from the edge of the one below (see also half bond).

SIDE-LOCK Tile longitudinal interlocking system, consisting of ribs and grooves to improve the resistance to water ingress.

SINGLE-CAMBER A traditional plain tile, arched along its length from head to tail providing a neat, clean aspect with emphasis on the course line of each row of tiles.

SINGLE-LAP Description for roofing materials such as concrete interlocking tiles and fixed-gauge clay tiles that rely upon an interlock at the sides of the tile to provide waterproofing (see also double-lap).

SINGLE-LAPPED TILES Tiles that have side interlocks and drainage channels to drain away the water.

SOAKER A small piece of sheet (usually lead), shaped and inserted between double-lap tile or slates on the abutment between a roof slope and a vertical wall, or at a mitred hip or valley.

SOAKER (DRY) A preformed component inserted between tiles or slates on the abutment between a roof slope and a vertical wall.

SOFFIT BOARD Board fixed to the feet of rafters which forms the underside of projecting eaves.

SOLDIER COURSE A decorative gable end finish normally done using plain tiles.

SPAN The distance between the walls which support the roof.

SPROCKET An alteration in pitch from steep to shallow, normally close to the eaves (Bell-cast).

STANDARDS Include international (ISO), European (EN) and UK (BS) standards that define recommendations for materials, design and workmanship used to ensure best practice.

SUSSEX CUT A decorative gable end finish normally done using plain tiles and tile-and-a-halves.

TACTRAY Structural metal liner system that replaces the need for timber rafters in steel-framed buildings

TILE-AND-A-HALF (gable tile) A tile, one and a half times the width of a standard plain tile, to maintain a broken bond at verges and abutments. Sometimes called a tile-and-half or a gable tile.

TILT Tile lift provided to ensure successive tiles lie correctly with no gap at their lower edge, and to direct water.

TILTING FILLET A tapered section of timber fitted to the eaves to support the first course of tiles or slates and prevent a trough in the underlay.

TRUSS A factory-made roof frame.

UNDERCLOAK Roofing slates, plain tiles or fibre cement strip fixed at the verge beneath the battens, which supports mortar on to which the verge tiles are bedded.

UNDER EAVES COURSE A row of shorter plain tiles or slates laid broken/half-bond under the first full course to maintain the required head and side laps.

UNDERLAY A breathable or non-breathable membrane acting as a barrier between the roof covering and the substructure (also called membrane or felt).

UNDERLAY SUPPORT TRAYS A preformed tray used to support the underlay at the eaves to prevent troughs and water traps.

VARIABLE-GAUGE TILES Tiles where the batten gauge can be easily adjusted (reduced) to fit the rafter length.

VALLEY The junction of two inclined roof surfaces at an internal angle.

VALLEY TILE A concrete or clay tile used at valleys with plain tiles.

VALLEY TROUGH A concrete tile fitting used for weathering valley junctions when using interlocking tiles (now superseded by preformed valley liners).

VAPOUR CONTROL LAYER (VCL) A non-permeable membrane fitted to the warm side of the insulation, intended to restrict the transmission of water vapour.

VENT TERMINAL A roof tile fitted with a hood and grille for natural ventilation which can also be connected to soil pipes or mechanical extractors.

VENTILATION TILE to ventilate the roof space or for soil vent pipe (SVP) or mechanical extraction.

Cowl Ventilators A cowl tile ventilator has a cowl or cap on the top of the ventilator (sticks out above the roof line).

Concealed Ventilators A concealed tile ventilator has an in-built grill which is inset into the ventilator (it does not stick out above the roof line).

VERGE A free end of a roof surface; for example, that at the end of a gable or dormer.

VERGE TILE A special tile to allow courses to be laid ‘broken bonded’ or special tiles designed to neatly finish off the left-hand side of the tiling so it is in keeping with the rest of the roof (i.e. double roll pan tiles).

VERTICAL TILING Sometimes known as ‘tile hanging’, where roofing battens are fixed to a vertical surface such as a wall or dormer cheek. Plain tiles are then fixed to the battens.

WALL PLATE Length of timber which is fixed to the top of a wall to secure the rafters.

WARM ROOF A roof designed with the insulation at rafter level.

WELL-SEALED CEILING Ceiling made to avoid constructional gaps and typically has an air permeability of not more than 30 mm^2/m^2.

WELT Lead flashings can have straight edges or have a welt which is a folded edge that strengthens the flashing.

WINCHESTER CUT A decorative gable-end finish normally done using plain tiles and tile-and-a-halves.

Key standards

This book contains references to many standards, codes of practice, technical documents and UK legislation concerning the manufacture and installation of roof tiles and associated roofing components. These are identified in the *Bibliography* and readers of this book are recommended to use the latest version of any standard or technical document referred to.

The three main standards relating to the design and installation of pitched roofs are BS 5534, BS 8000–6 and BS 5250. As of 1st January 2024, the latest version of each is:

- BS 5534:2014+A2:2018
- BS 8000–6:2023
- BS 5250:2021

BS 5534 *Slating and tiling for pitched roofs and vertical cladding. Code of practice*.

BS 5354 gives recommendations primarily intended for the design, performance and installation of new-build pitched roofs, plus vertical cladding. It looks at normal reroofing work, both new installations and repairs, using clay tiles, concrete tiles and slates, all with their associated components. It does not cover the structural design of a roof. As a code of practice, its suggestions are not a legal requirement, but a roof specified to BS 5334 can be upheld by law.

BS 8000–6 *Workmanship on construction sites – Slating and tiling of roofs and walls. Code of practice.*

The BS 8000 series of standards was created with the aim of raising end quality and performance in the built environment. The standards cover a range of topics, including safety, health and welfare, materials, workmanship and the management of construction projects. They provide guidance on how to ensure that the work is conducted to a high standard and meets the requirements of the client.

Part 6 gives recommendations on basic workmanship on building sites, both commercial and domestic. It covers those tasks which are frequently used in the tiling of roofs and walls, both new builds and refurbishments.

BS 5250 *Management of moisture in buildings. Code of practice*.

BS 5250 gives guidance on managing the risks associated with excessive moisture within buildings, such as mould growth, building structure failures and occupant health problems. It outlines the primary origins of water vapour, its movement and deposition, offering advice on mitigating these hazards throughout the design, building and occupation of buildings.

Full copies of all standards can be obtained from the BSI (British Standards Institution).

CHAPTER 2

Design and performance

2.1 Common design considerations

The design issues that need to be considered during the design stage of a pitched roof can be divided into 'prescriptive design specifications' and 'performance design specifications'.

Prescriptive design specifications

Rafter/pitch of the roof

Plain tiles

Clay or concrete plain tiles conforming to the dimensional tolerances given in BS EN 1304 or BS EN 490 can be laid on rafter pitches down to a minimum of 35°. Specialist tiles that are close fitting when laid may be used on roof pitches below 35°, provided the manufacturer is able to produce evidence of satisfactory prolonged and extensive use of the product, including at the roof pitches, head-laps and side-laps for which they are intended to be used.

Plain tiles and peg tiles that, for aesthetic reasons, do not comply with the dimensional tolerances given in BS EN 1304 must be laid at pitches not less than 40°.

Interlocking tiles

Clay and concrete interlocking tiles conforming to the dimensional tolerances given in BS EN 1304 or BS EN 490 and when without effective anti-capillary devices can be laid on rafter pitches down to a minimum of 30°. Proprietary tile designs that have effective anti-capillary devices, head-lap and side-lock features, and are close fitting, may be used at pitches below 30° provided that evidence is available from the manufacturer of satisfactory prolonged and extensive use of the product (a minimum of 15 years) at the roof pitches, head-laps and side-laps for which they are intended to be used.

Where a low-pitch roof is required, manufacturers should provide evidence based on an appropriate test method that can be directly correlated with the recommended conditions of use, for example PD CEN/TR 15601 *Hygrothermal performance of buildings. Resistance to wind-driven rain of roof coverings with discontinuously laid small elements – Test methods*.

DOI: 10.1201/9781003196990-2

Head-lap and side-lap

Plain tiles

For plain tiles on pitched roofs, BS 5534 recommends a minimum head-lap of 65 mm, which, translates to a maximum batten gauge of 100 mm. Gauges of less than 88 mm are not recommended. The side-lap should be not less than one-third the width of the tile, typically 55 mm.

For vertical walls (over 70° pitch), BS 5534 recommends a minimum head-lap of 35 mm, which translates to a maximum batten gauge of 115 mm. The side-lap should be not less than one-third the width of the tile, typically 55 mm.

Interlocking tiles

For pitched roofs, the head-lap specification in BS 5534 for variable gauge tiles without a headlock is 75 mm minimum. The head-lap and side-lap recommendations of products that have a design feature such as anti-capillary grooves on the top of the underside surfaces of the tile and/or a side-lock are determined by the design and subject to manufacturers' recommendations.

For vertical walls, the head-lap is determined by design features on the tile face at the head of the tile, and/or features on the underside at the tail of the tile, and subject to the manufacturer's recommendations. In BS 5534 it is recommended that for variable gauge products, the head-lap should be not less than 35 mm below the nail hole, if present.

Performance design specifications

Wind load

On the leeward (downwind) side of a building the wind can create a suction on the tiles and the uplift effect can be significantly higher on those tiles adjacent to the perimeters. The methods for calculating the wind uplift load are given in Chapter 2.3 (of this book) *Wind uplift calculation* and in BS 5534.

The minimum fixing specification for plain tiles with nibs laid on roof pitches below 60° is to fix every fifth row with two nails in each tile. For roof pitches of 60° and above, including vertical, two nails should be used to fix every tile. At verges and abutments, and at each side of valleys and hips, the end tile in every course should be twice nailed where possible or mechanically fixed. At eaves and top edges, two courses of tiles (consisting of under eaves courses, first full course at eaves, last full course and tops course at top edges) should be nailed twice or mechanically fixed.

Traditionally plain tile peg tiles are not nailed; rather, they are once pegged to allow them to be aligned in horizontal coursing.

The minimum fixing specification for interlocking tiles is to mechanically fix each tile. For rafter pitches of 45° and over, each tile should be nailed with at least one nail. Additionally, for rafter pitches of 55° and over, including vertical, the tail of each tile should be mechanically fixed, meaning clipped. At verges, abutments, each side of valleys and hips, and eaves and top edges, perimeter tiles for these details should be twice nailed or mechanically fixed.

Depending on the outcome of wind uplift calculations, it may be necessary to consult the tile manufacturer who will provide a site-specific specification including the use of fixing methods such as ring-shanked nails, screws, clips or proprietary fixings.

Note: The Building Research Establishment (BRE) has published a guide, Digest 467 *Slate and tile roofs: avoiding damage from aircraft wake vortices*, that describes the effect of aircraft vortices on roofs and gives recommendations for the fixing of tiles in areas that are on the flight path of aircraft taking off and landing.

Ventilation

Poorly designed roofs with inadequate ventilation can lead to significant problems due to high levels of moisture in the building and therefore, risk of condensation. BS 5250 was revised in 2021 to provide moisture management guidance broadened to cover not only condensation but also other sources of moisture such as wind-driven rain and rising damp. BS 5250 contains guidance for how to manage condensation risk in pitched roofs depending on the type of pitch roof construction. Normally, this will involve ventilating the cold roof spaces wherever they occur. The method of ventilation should therefore be established prior to the assembly of the roof covering.

The position of roof insulation will affect the choice and method of ventilation used, and Chapter 2.10 (of this book) *Ventilation in warm and cold roofs,* describes examples where the insulation is positioned horizontally at ceiling height (cold roof) or at pitched rafter level (warm roof). The ventilation method adopted must also take into consideration the vapour resistance of the underlay that will be used. BS 5250 defines the following:

- Airtight layer – prevents the movement of air which may/may not act as a vapour control layer.
- Air & Vapour Control Layer – a material which can limit both movements of vapour by diffusion, and air movement (sometimes called vapour checks or vapour barriers).
- High Resistance (HR) underlay – an underlay with a vapour resistance greater than 0.25 MNs/g.
- Low Resistance (LR) underlay – an underlay with a vapour resistance less than 0.25 MNs/g (sometimes referred to as vapour permeable or vapour open underlays, and some may possess a degree of air permeability).
- Breather Membrane – defined as a membrane with a vapour resistance less than 0.6 MNs/g (sometimes this terminology is used, arguably wrongly in place of LR underlays).

The method of assessment given in BS 5250 should be used and, where the risk of condensation is identified, appropriate ventilation to the roof spaces or voids should be provided and/or an AVCL should be incorporated within the ceiling/wall structure in order to prevent water vapour from reaching the cold side of the insulation. See Chapter 2.9 *Ventilation in warm and cold roofs*.

Rain and snow resistance

The resistance to rain and snow is largely determined by the lap arrangement. The lap arrangement described in BS 5534 for plain, interlocking, clay and concrete tiles provides an excellent rain and snow protection system and is described further in Chapter 2.5 *Rain and snow resistance*.

Tile durability

Tiles that meet the stringent requirements of BS EN 1304 or BS EN 490 have demonstrated that they have the necessary durability for the UK environment. Further detail can be found in Chapter 2.2 *Material specification*.

Thermal capacity

Although the thermal capacity of clay or concrete tiles can help to regulate the internal temperature of a building, the thermal resistance properties of clay and concrete tiles can be ignored as they do not, in and of themselves, provide a significant level of insulation. Roofing tile properties are specifically related to water permeability, durability and aesthetics. It is the role of the insulation and other thermally efficient products to provide the necessary thermal performance of the building structure.

Fire resistance

The reaction to fire of a roofing or cladding product and its assembly in terms of flame spread and flame penetration should be determined by reference to the properties of the roofing or cladding element, the method of assembly, the effects of the sub-elements and the air permeability of the array as laid.

Both clay and non-coated concrete tiles manufactured to BS EN 1304 and BS EN 490 used on a conventional pitched roof with insulation, underlay, battens and tiles, are deemed to satisfy the respective national building regulations in the UK with respect to external fire performance and are designated non-combustible. In recent years, fire spreading between homes and multiple occupancy buildings has caused a change of focus regarding fire safety and the use of cavity barriers and/or fire stops.

Vertically tiled or slated walls have concealed spaces with paths that can allow smoke or flame to enter the batten cavity and should be sealed using a suitable timber batten with a minimum 38 mm thickness, steel of a minimum 0.5 mm thickness or a manufactured cavity barrier. Cavity barriers should be positioned to close off penetrations through the wall such as windows, doors and openings, and on the line of any compartment wall or floor. Both vertical closed-state cavity barriers and horizontal open-state cavity barriers should be installed to balance the requirement for the cavity to be ventilated to allow air movement for the removal of moisture and provide drainage with the need to seal the batten cavity to prevent the spread of flames, heat and smoke during a fire.

All vertical closed-state cavity barriers should be installed prior to the horizontal cavity barriers, and when under compression, the vertical cavity barrier should not lift the tiles or slates. Horizontal cavity barriers should be installed with the intumescent front face on the leading edge. The manufacturer of the cavity barrier should be consulted regarding the width of the cavity so that the correct size of cavity barrier is sourced prior to commencing any installation.

The England Building Regulations Approved Document B and/or its national equivalent in Scotland, Wales and Northern Ireland should be consulted for full details.

Insects, birds and bats

The correct design and installation of a tiled roof will ensure that the ingress of insects and birds to the loft or wall structure is prevented.

Protected species, such as bats and swifts, that often nest in roof spaces, can be accommodated by the use of special roof fittings that provide an entry and exit to nesting located within the roof space. Some manufacturers offer such products for this purpose.

Furthermore, when bats use the roof as a roost the choice of roofing underlay can be restricted as some types of underlay have been shown to cause them harm. Non-bitumen-coated roofing membranes currently are underlays of a spun polyester-bonded construction made up of long fibres. Bats' claws are sharp enough to pull these fibres loose, which can entangle and ultimately kill.

A bitumen 1F membrane that is of non-woven short-fibre construction is considered to be a bat-safe roofing membrane and does not need a certificate for Building Regulations licence applications.

Some manufacturers are formulating Non-Bitumen Coated Roofing Membranes (NBCRM's) which will be alternative 'bat safe' options. The underlay manufacturer or specialist bat experts should be contacted for guidance.

2.2 Material specification

Construction products are an increasingly well-regulated sector with statutory and voluntary schemes available to confirm compliance. This chapter confirms the relevant standards and conformity scheme required for roofing products.

Tiles and fittings

Plain and interlocking **clay** tiles and fittings should comply with BS EN 1304.

Plain and interlocking **concrete** tiles and fittings should comply with both BS EN 490 and also BS EN 491.

Accessories

Dry fixed hip, ridge and verge systems should comply with BS 8612. Products not covered by this standard such as ventilation tiles, proprietary soakers, outlets, proprietary flashings, etc. should have the United Kingdom Accreditation Service (UKAS) Accredited Third Party Certification or the product's performance should be demonstrated based on evidence of successful long-term history in use.

Lead

Where lead is used for a valley or gutter lining, flashing or saddle exposed to weathering it should be minimum Code 4 in accordance with BS EN 12588, and where it is used as a protected soaker beneath tiles it should be Code 3. Surfaces of all exposed lead should be treated with patination oil to prevent lead oxide staining of the tiles. Further details can be obtained from the Lead Sheet Training Academy.

Mortar

Where mortar is used as a filler, a mix of one part Portland cement to three parts sand mix (with the sharp sand making up no less than one-third of the sand content) with plasticiser in accordance with the manufacturer's instructions is recommended. Most sands conforming to BS EN 13139 are suitable. Where mortar is used to bed hip, ridge, valley or verge tiles, the tiles must also be mechanically fixed, with the exception of baby ridge and hip tiles at low level, subject to site wind uplift load requirements. Specific lime mortars may be required for historic or listed buildings. Guidance on roof-tile mortar mixes is given in BS 5534.

Adhesives

Where adhesives are used, the adhesive manufacturer's recommendations should be followed to ensure that the product is suitable for securing concrete or clay tiles. The use of suitable adhesives should only be considered in circumstances where tiles and fittings cannot be mechanically fixed, such as small cut tiles which are less than half a tile width for example.

Underlays

Fully supported

This includes roofing underlays laid directly onto a rigid boarding, insulation or sarking. The roofing underlay should be of adequate strength, water resistance and durability with water vapour transmission high enough to prevent the formation of condensation beneath the underlay. The method of assessment given in BS 5250 should be used to ensure that harmful condensation will not develop. If necessary, to overcome potential condensation risks, a vapour control layer should be incorporated within the structure.

Unsupported

This includes roofing underlays which are draped over the rafters or underlays laid over counter battens on boarding, insulation or sarking. Roofing underlay should be of adequate strength, water resistance and durability in accordance with the requirements of BS 5534 and BS EN 13859–1. Reinforced bitumen underlays should conform to the requirements of BS 8747 for type 1F or 5U.

Flexible underlays

Reinforced bitumen underlay of type 1F or 5U that meets the requirements of BS 8747 or non-bituminous flexible underlay that meets the requirements of BS EN 13859–1 should be used. Products not meeting these standards should have a UKAS Accredited Third Party Certificate appropriate for the conditions of use.

Rigid underlays

Rigid underlays comprising flat or profiled sheets, wood-based panels, fibre cement sheets, corrugated bitumen sheets or sheets of other materials should conform to BS EN 14964. Products that do not meet the requirements of BS EN 14964 may be used provided there is evidence they are fit for purpose and have proven relevant experience in practice or have test method data based on UK conditions and methods of use.

Battens and counter battens

BS 5534 sets out a number of requirements relating to battens and counter battens. In summary, these are:

- Timber type.
- Timber quality.
- Batten's preservatives.
- Batten sizes.
- Batten fixings.

Chapter 2.6 *Battens* gives more detail about battens' requirements.

Fasteners

Nails for tiles

Generally, clout head nails of diameter 3.35 mm and 2.65 mm are used for most applications (subject to withdrawal resistance determined by wind uplift calculations). Tile-fixing nails should penetrate the batten by a minimum of 15 mm. Clout head nails complying with BS 1202 part 2 (copper), part 3 (aluminium), BS EN 10088–3 grade 304 or 316 (stainless steel) or BS EN 10230–1 (zinc-coated steel) may be used. Nails made of other materials, including phosphor or silicon-bronze, may be used for situations with aggressive environmental conditions and should satisfy the conditions of the manufacturer's specification.

Aluminium is the most common material used for tile nails. Steel nails should not be used for nailing tiles and where galvanised, should only be considered where there is no risk of the protective coating being damaged during installation.

Stainless steel is often used where enhanced durability or strength is required. Improved nails (annular, ring shank and drive screws) conforming to BS EN 10230–1 and BS 1202–1 or screws may be used where the wind load calculation indicates that smooth shank nails will not meet the requirement.

Tile clips

Tile clips are made from a wide variety of materials. Manufacturers supply proprietary clips to suit their tiles, and it is important that they are used in conjunction with a site-specific fixing specification. Use of third party supplied clips may invalidate manufacturer warranties. Refer to BS 5534 for further information.

Fasteners for fittings

Hip irons are hooks that fix to the lower end of the hip rafter and provide mechanical security for the bottom hip tiles. All mortar bedded ridge, hip and verge tiles should be mechanically fixed. Manufacturers provide suitable fixings for this purpose. Alternatively, ridge, hip and verge components can be dry fixed using proprietary products that should meet the requirements of BS 8612.

Nails and fasteners for securing battens or counter battens to rafters

Round wire nails complying with BS EN 10230–1 galvanised or BS EN 10088–3 grade 304 or 316 stainless steel should be used. The nails should generally be not less than 3.35 mm in diameter and 65 mm long, and provide a minimum 40mm penetration into the rafter, however, they may be longer to meet the requirements for wind loading. Where proprietary nail fixings are to be applied using mechanical nail guns, designers and specifiers are advised to seek guidance from the manufacturer as to their suitability for the intended use. For exposed conditions, improved nails, screws or helical fixings may be required. BS 5534 Annex H7 gives guidance on calculating the required wind load resistance for fixings used to secure battens and counter battens.

Flashing and junctions

Where required, metal flashings and junctions should be fixed with copper or stainless-steel nails. The size of the nails should be in accordance with the recommendations of the Lead Sheet Training Academy (LSTA). Aluminium nails must not be used to fix lead flashings. Flashings in exposed locations may need to be clipped, and this should be in accordance with the Rolled Lead Sheet Manual recommendations.

2.3 Wind uplift calculation

When designing a new roof, wind load calculations are a key part of the process. Wind load calculations involve calculating the potential wind uplift to inform the design of the roof so that it can resist the potential uplift. A roof that is not prepared for a strong wind uplift will be damaged, and this will lead to expensive repairs.

Wind uplift is caused by gusts of wind hitting the side of any building. This disturbance in the wind movement can direct it upwards at an increased speed. Once the wind reaches the top of the building it can continue to move freely, but before it does so it forms vortices. These cause areas of negative pressure that can settle above the roof. The negative pressure creates an uplift force that has the power to pull roof coverings completely off the structure.

There are different factors that will contribute to the effect of the wind uplift, and these include:

- the location of the building, immediate surroundings and geographic location
- the height of the building
- the width and length of the building
- the roof shapes
- the roof pitches
- the tile types

During the design stage, specific wind uplift calculation incorporating the full requirements of BS 5534 should be carried out. The detailed calculations to determine wind uplift can be found in BS 5534.

The following sections outline the various factors that resist wind uplift and the minimum requirements for fixings. Further advice on fixing specification and wind uplift are available from the tile manufacturer.

Mass resistance of roofing products

The mass resistance against uplift and overturning should be taken as 0.9 times the average mass of the roofing element when there are no fixings present. When fixings are used in conjunction with the roofing element, the mass resistance against uplift and overturning can be taken as 1.0 times the average mass of the roofing element.

Note: Mass resistance equates to the dead load or weight of the roof-tile covering.

Resistance of roofing underlay and board sarking

Underlays and board sarking should have adequate resistance and stiffness to resist wind loads. The upward deflection of a flexible underlay under maximum wind uplift load with battens at the maximum design gauge should be such as to avoid contact with the underside of the slates or tiles.

The resistance of the underlay should be calculated for the actual site-specific wind loading conditions using the full BS 5534 wind uplift requirements, including the condition of the ceiling (if fitted).

The UK has five wind zones that indicate the strength of the wind in each given area. Always check with the manufacturer to make sure the underlay can be used in a particular zone. Wind zone suitability applies to underlays where:

- there is a continuous ceiling
- the ridge height is not greater than 15 m
- the roof pitch is between 12.5° and 75°
- the site altitude is not greater than 100 m
- the site is not more than halfway up a hill or escarpment with more than 5% gradient

Resistance of mechanical fixings

The resistance to wind uplift of nails, screws and mechanical fixings such as clips and hooks connected to the roofing substrate can be determined by the test methods given in BS 5534.

Resistance of mortar

Any tensile bond strength offered by the mortar bedding of concrete or clay ridge and ridged hip tiles, verge and valley tiles should not be taken into account when determining the mechanical resistance required to resist wind uplift. All mortar bedded components should be supplemented by mechanical fixings to secure them to the roof structure where the wind uplift load exceeds the dead load resistance.

Combined resistance

In calculating the combined resistance to uplift, the resistance due to the mass of the roofing element, for example ridge/hip/verge/valley, may be added to the resistance of the mechanical fixings used to secure such elements, but no account should be taken of the tensile strength of the mortar.

The following section sets out the minimum recommendations for mechanical fixings.

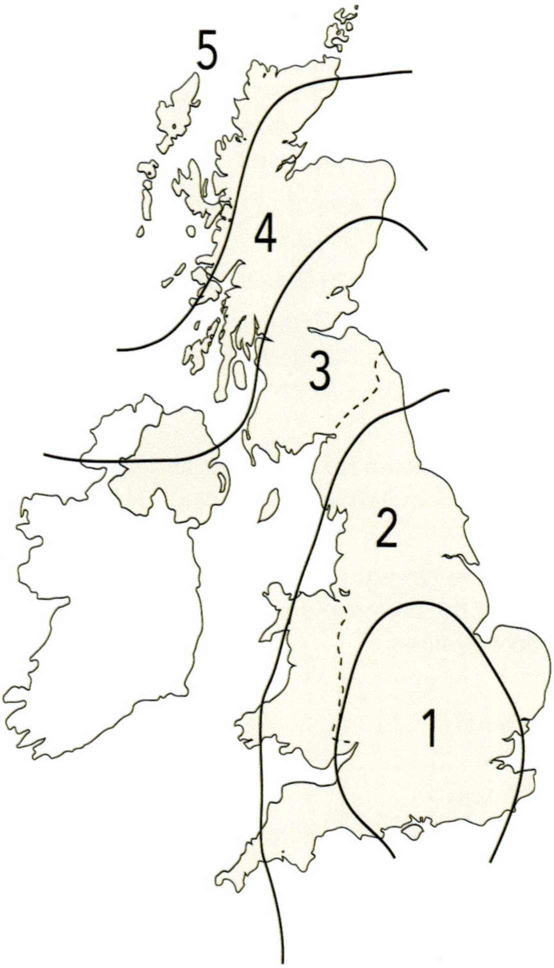

BRE wind zone map

Geological wind zone	Design wind pressure, p_u, for underlay (N/m^2)
5	1600
4	1330
3	1150
2	1150
1	820

BRE Digest 489 (8)

Figure 2.1 Wind zone map

General

When determining the fixings for roofing elements, the following recommendations should be met:

- The perimeter roof cladding elements should be mechanically fixed using a minimum of two fixings (subject to meeting the wind loading recommendations), one of which can be a tile clip, adhesive or dry verge capping system where appropriate. Perimeter means the single element at any discontinuity in the plane of the roof, including roof edges. To avoid the use of small pieces of cut tiles, which are difficult to fix, double tiles, tile-and-a-half or half tiles should be used where available from the manufacturer. Small pieces (less than half tile width) of cut single-lap tiles should be bonded or mechanically fixed to the adjoining full width tile.
- Both of the following two points should be taken into account:

1) Recommendations for general security against dislodgement, for example due to vibration or access traffic.
2) Fixing recommendations appropriate for the calculated wind loads. Care should be taken to ensure that the necessary fixing penetration is achieved where the insulation elements are located above the rafter level.

Double-lap tiles

For nibless tiles, two nails/pegs should be used to every tile. Where nailed, plain tiles should have at least the following fixings:

- For nibbed tiles, for rafter pitches below 60°, two nails should be used in each tile in at least every fifth course. For rafter pitches of 60° and above, including vertical, two nails should be used in every tile.
- At verges and abutments, and at each side of valleys and hips, the end tile in every course should be twice nailed (mechanically fixed) where possible.
- At eaves and top edges, two courses of tiles (consisting of under eaves course, first full course at eaves, last full course, and tops course at top edges) should be nailed or otherwise mechanically fixed.

Nails for plain tiles should be not less than 2.65 mm diameter and should penetrate the battens not less than 15 mm. Peg tiles and nibless tiles require special treatment, and you should consult Chapter 7.2 *Historic heritage roofing – Laying of historic peg tiles*.

Single-lap tiles

Single-lap tiles should have at least the following fixings:

- For all roof areas and rafter pitches, every tile should be mechanically fixed with at least one fixing per tile as a minimum.
- For rafter pitches of 45° and over, each tile should be nailed with at least one nail. Additionally, for rafter pitches of 55° and over, including vertical, the tail of each tile should be mechanically fixed.
- At verges and abutments, and at each side of valleys and hips, the end tile in every course should be twice mechanically fixed.
- At eaves and top edges, one course of tiles should be twice mechanically fixed.

Nails for single-lap tiles should be of a length to provide adequate resistance to wind uplift, subject to a minimum penetration of 15 mm. The presence of nail holes reduces the effective lap of some patterns of tiles. Accordingly, tiles for low pitches are often designed for fixing by nibs and clips. Such clips are usually fixed at or near the tail of the tile and connected to battens or boarding. Where a tile has two nail holes and is fixed with a single nail, the nail hole nearest to the overlock (typically the right-hand nail hole) should be used.

Underlays, sarking and battens

In addition to checking for resistance to wind uplift, the nailing for sarking boards and battens for general security should be at least one round steel-wire nail (or two nails in the case of a butt joint) of 3.35 mm diameter and a penetration of not less than 40 mm.

Battens and counter battens fixed to masonry should be plugged or shot-fired with connections of proven adequate pull-out resistance. Cut nails driven into masonry should only be used when loaded in shear and not when loaded in withdrawal.

Nails for use with roofing underlays should be clout head nails of not less than 3.0 mm shank diameter and 20 mm length. Proprietary fixings for use with sarking boards, underlays, battens and counter battens should be durable and capable of resisting the dead loads, snow loads, wind loads and imposed loads applicable to the roof and its location.

Flashings

Preformed flashings should be installed in accordance with the manufacturer's recommendations and rolled sheet flashings in accordance with the Lead Sheet Manual.

Fittings and accessories

In addition to checking for resistance to wind uplift and differential movement, the fixings used for both mortar bedded and dry fixed ridges and hips should be of proven durability and used in accordance with the manufacturer's recommendations.

Spacing of nails and screws

To minimise timber splitting, nails and screws in timber-to-timber joints should be positioned with the minimum spacing as a multiple of the nail diameter or screw diameter.

BS EN 1995–1–1 Eurocode 5 contains details and design rules for timber connections using metal fasteners.

Clips

All clips should have adequate resistance to the calculated wind uplift loads and should allow the roofing element to return to its laid position after maximum wind load events.

2.4 Roof pitch

Pitched roofs come in all shapes and sizes, from simple designs to complex shapes. Roof shape help defines the overall feel of a building, with steeply pitched roofs offering a more traditional look versus the more contemporary aesthetic of a low or asymmetric pitch.

Modern pitched roofing trends are being influenced by the fact that owners are increasingly likely to invest in developing their property. Buildings with steeper roof pitches allow for the creation of habitable spaces by converting loft areas, whilst ground floor extensions often require lower pitch roofs to avoid blocking first-floor fenestration. Planning regulations allow for domestic extensions up to a certain size to be added to properties without needing planning permission (permitted development), so manufacturers have developed low-pitch products to provide roofing solutions to this market.

An optimum range for roof pitches in the UK, with its often wet and windy weather conditions, is between 30° to 50°. A roof pitch in this range provides protection from the elements, effective water drainage, peak thermal performance and good ventilation, helping to create a comfortable environment throughout the changing seasons. However, with projects such as domestic extensions, this pitch angle is not always possible.

An effective alternative to a flat roof, low-pitch roofs are designed so that water runs off safely without ponding and finding its way into the building structure. Generally, double-lap plain tiles should be laid not less than 35°, and single-lap tiles no less than 30°. Where the roof pitch is below 35° there are restrictions on the type of roof coverings that can be used but lower pitches are possible if specialist products are used. See section 2.1 *Common design considerations* for more details.

When opting for a low-pitch roof design, several important considerations need to be made. At a low pitch, some elements of the roof system are working at the limit of their capabilities, meaning that there is a higher chance of failure under extreme weather conditions. Although individual roof components have been thoroughly tested and perform perfectly well at the given pitch, constructing the roof with the same accuracy that is used under test conditions takes skill. It only takes a small gap or two in the tiling where it passes over a flashing and the risk of water ingress from rainfall or wind-driven rain is increased.

In terms of effective water drainage, a roof with a shallower slope discharges rainwater at a slower rate than a steeper roof, simply due to gravity. When a roof structure also features long rafters, this can mean that a large amount of water can remain on the surface and find its way into any gaps, damaging underlay and battens over a prolonged period. Therefore, roof-tile manufacturers often recommend a limited maximum rafter length on low-pitched roofs. The general 'rule of thumb' advice is that pitch should be increased 1° for every half-metre distance that the rafter length is over the recommended maximum length, but as this may vary depending on the products used, the specific project criteria and exposure, guidance should always be obtained from the manufacturer.

During rainfall, water runs down tiles to the courses below, and it is the sufficient side-lap and pitch which prevents sideways or upwards creeping of water. It is also important to pay

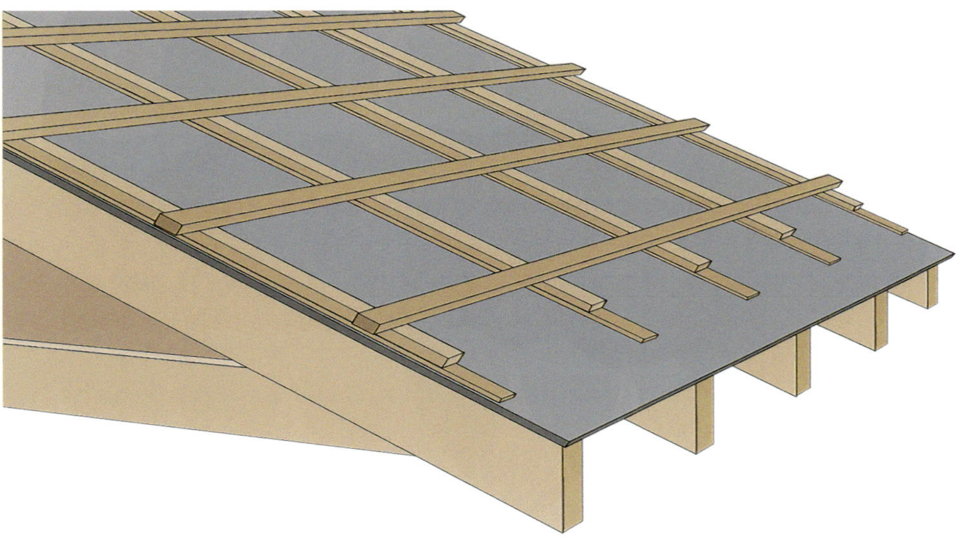

Figure 2.2 Roof specification must be appropriate for the roof pitch

close attention to the recommended head-lap stated on the manufacturer's instructions, as often the lower the pitch, the larger the head-lap should be.

Low-pitched roofs constructed on single storey extensions may also have to cater for rainwater being discharged from larger roof slopes above. In this circumstance, the rainwater pipe should be directed to a purpose-made gutter trough formed in the lower roof slope, rather than be allowed to surcharge onto the tiles. Therefore, careful design is important, with such point loads avoided if possible.

Rather than relying on standard roofing construction, which is fine for steeper roofs, with low pitches it is advisable to detail and construct the roof to protect vulnerable areas such as eaves, valleys, abutments, roof window surrounds and other such junctions. These areas need to be considered even more carefully with low-pitched structures to make sure that they are contributing to a weather tight roofing system.

2.5 Rain and snow resistance

General

The UK has a climate where there is a high risk of severe wind-driven rain. Roofing products, fittings and accessories, when laid and fixed to a roof, perform in different ways to resist snow and rainwater penetration.

The mechanisms of rainwater ingress with roofing products are many and varied and can include:

- capillary action and water creep
- driving rain
- deluge rain and flooding
- raindrop bounce
- additional water on roofs with long rafter lengths
- rainwater discharge from an adjacent roof onto the roof covering
- wind-driven snow
- water displaced around roof windows

The extent to which a roofing product is susceptible to weather ingress depends on the design details, pitch, lap, gaps as laid and geometry of the products in relation to the roof, the design, orientation, location and geometry, plus the macroclimate and microclimate affecting the roof.

A map of the UK details categories of exposure based on driving rain penetration data is given in BS 8104 and BRE Report 262. This map should be used when designing buildings up to 12 m ridge height. More detailed guidance is also given within BS 8104.

For buildings above 12 m in height, the influence of increased wind speed can be determined using BS EN 1991–1–4.

Recommendations found within BS 5534 for the performance of roof tiles against driving rain are based on real experience and performance over many years of use in all climatic locations.

PD CEN/TR 15601 may also be used as a comparative method for determining the performance of roofing tiles to resist driving rain when installed outside of the scope of experience quoted in BS 5534.

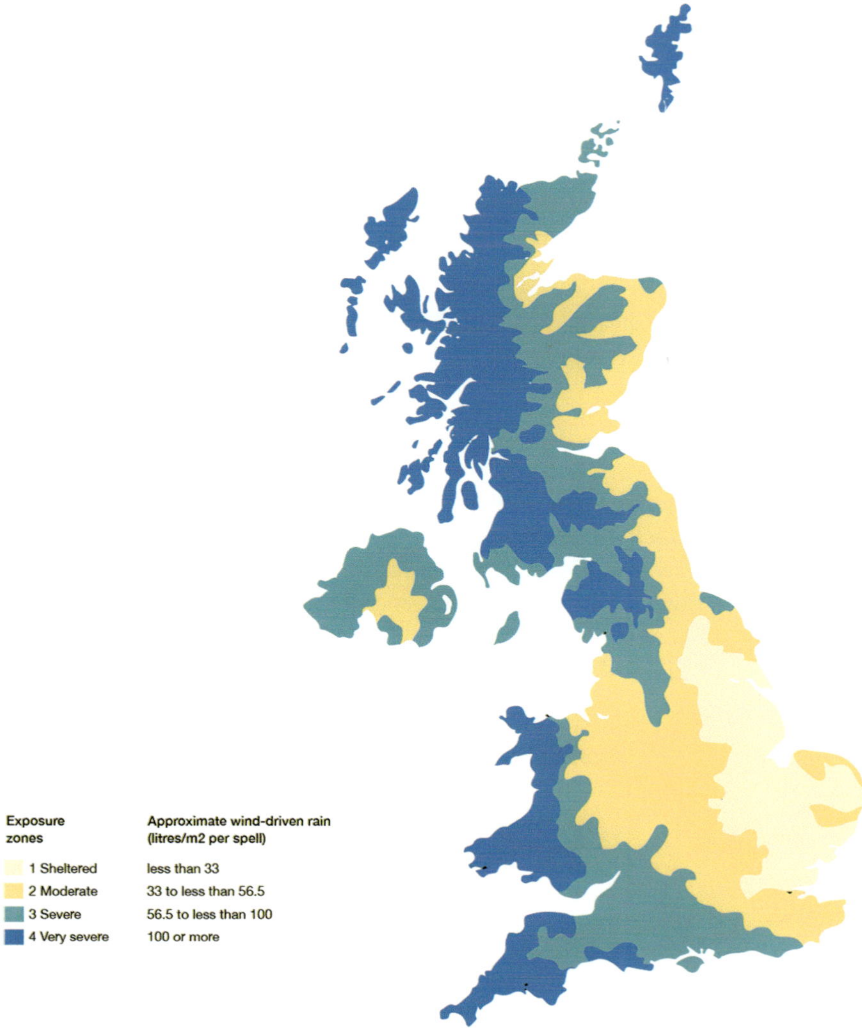

Figure 2.3 Exposure zone map (courtesy of the BRE)

Roof pitch, head-laps and side-laps

Once the exposure zone has been determined, it is necessary to finalise the roof pitch, which will inform the necessary head-lap and side-lap to achieve resistance against rain and snow. For calculating head-laps and side-laps, the roof pitch can be taken to be equal to the rafter pitch.

The actual pitch of a tile when laid will be slightly less than the roof pitch. This is additionally so when the tiles incorporate longitudinal camber.

Head-laps (double-lap)

The head-lap for double-lap products is the distance by which the upper course of the tile provides a lap with the next but one course below. Head-laps for double-lap tiles should be no less than that recommended in Chapter 2.1 *Common design considerations*.

Head-laps (single-lap)

Head-laps for single-lap tiles should also be no less than that recommended in Chapter 2.1 *Common design considerations*. Similarly, the head-lap for single-lap tiles is the distance by which a course of tiles provides an overlap with the next course below.

Some single-lap products have a head-lap of fixed dimension resulting from the design of the product.

Side-laps (double-lap)

For double-lap tiles, the notional side-lap is the side distance by which the tile overlaps the tile in the next course below. The side-lap for plain tiles at 100 mm gauge should be not less than 55 mm.

Side-laps (single-lap)

The side-lap for single-lap tiles should be in accordance with the minimum values shown in Chapter 2.1 *Common design considerations*. The side-lap for single-lap products is the amount by which one tile overlaps the adjacent tile in the same course by way of a side interlock or upstand feature. Single-lap interlocking tiles generally have a proprietary side-lock design.

Illustrative diagrams

The following diagrams are shown so as to clearly show the important positions of head-lap and side-lap.

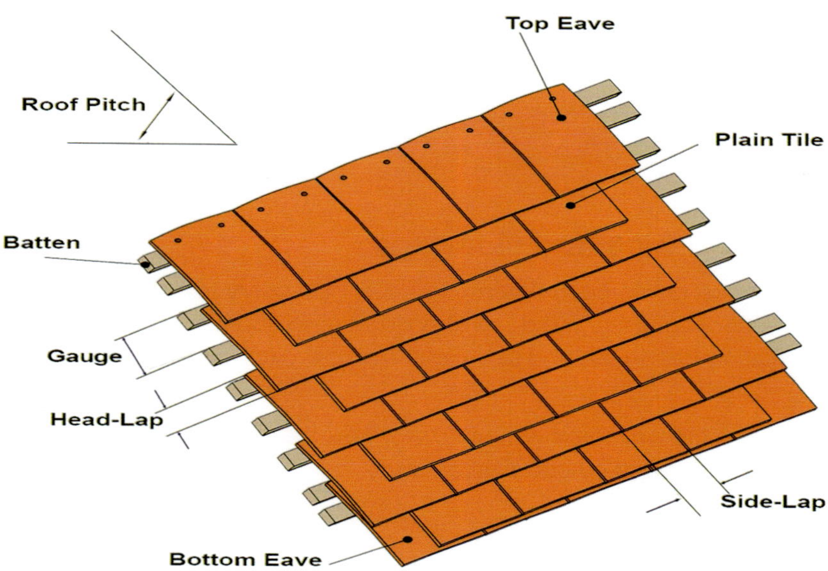

Figure 2.4 Side-lap and head-lap and pitch (face view – plain double-lap tiles)

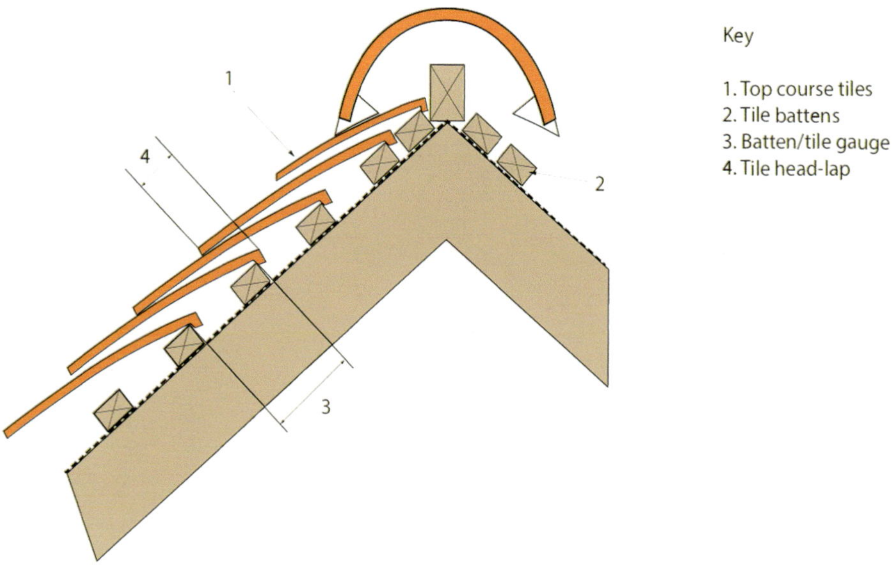

Figure 2.5 Head-lap and batten gauge (side view)

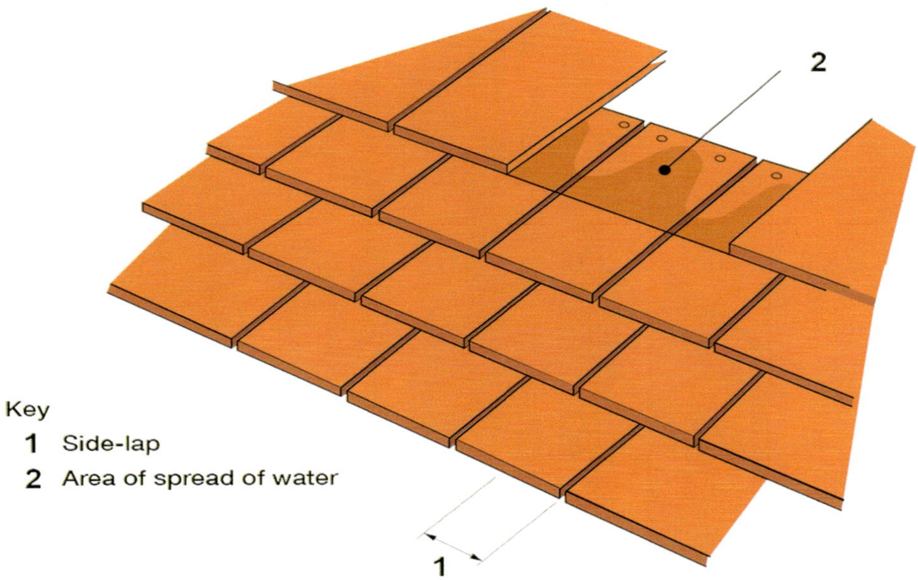

Figure 2.6 Side-lap and water spread

2.6 Battens

BS 5534 includes the specification for softwood timber battens used for tiles. It prescribes minimum sizes and tolerances, as well as the species to be used, and provides a detailed list of permissible characteristics and defects which battens must not exceed.

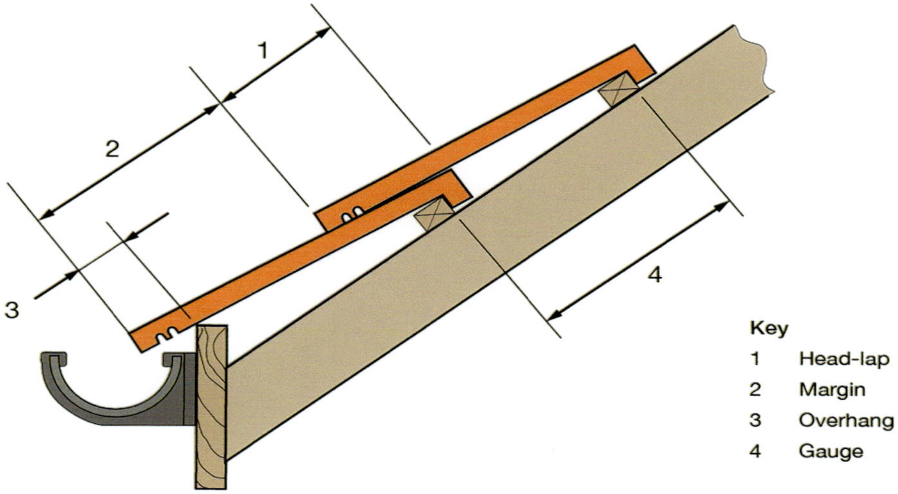

Figure 2.7 Lap and pitch of tiles in single-lap tiles

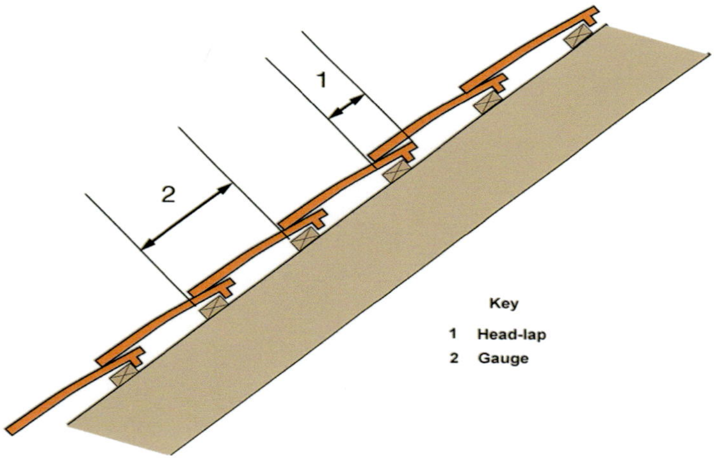

Figure 2.8 Single-lap tiles showing lap and batten gauge

Any claim of complete compliance with the recommendations of BS 5534 must incorporate the grading of battens. In 1997, a decision was taken, after wide consultation, that the requirements for the grading and quality of timber battens should be fully specified. This was necessary to deal more precisely with the small timber sizes not provided for in the then-current BS 4978.

Requirements

Reference should be made to BS 5534 for the provisions relating to the battens, BS 8000–6 for its guidance regarding the workmanship and installation of battens and to BS 8417 for its guidance relating to timber treatments.

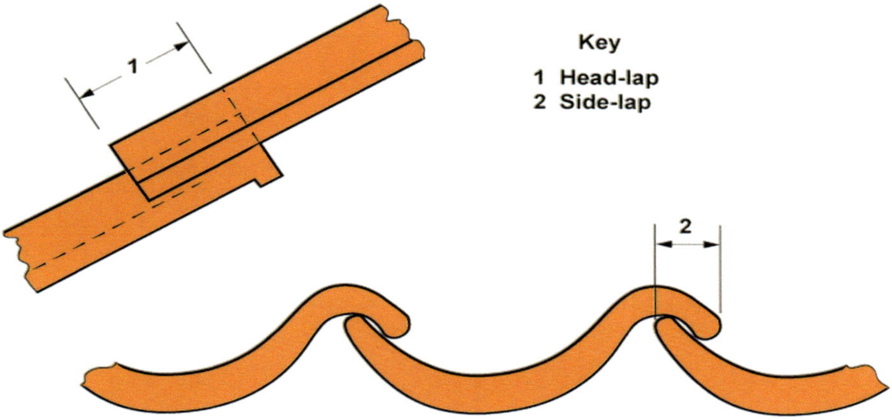

Figure 2.9 Side-laps and head-lap for single-lap tiles without interlocks

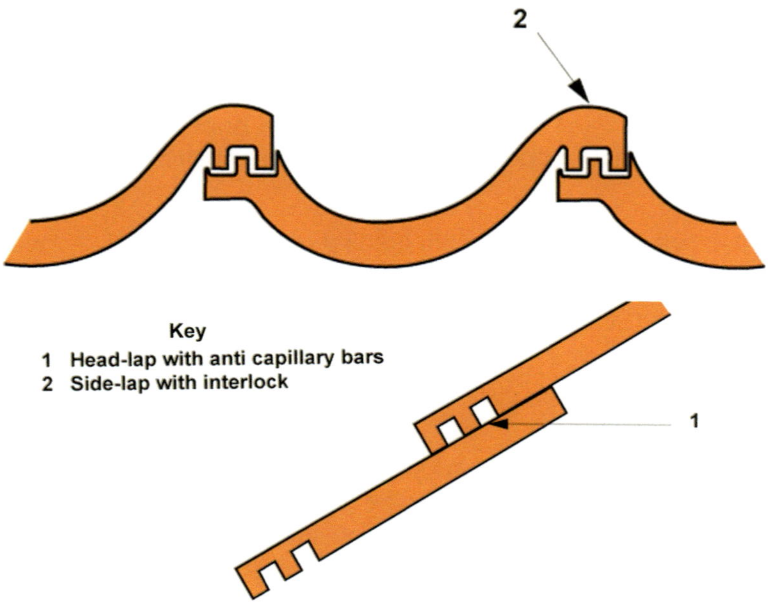

Figure 2.10 Side-laps and head-lap for single-lap tiles with interlocks

Timber species

The timber species should comply with type A or type B as specified in BS 5534 and should be preservative treated where the building regulations and by-laws require protection against the House Longhorn beetle. Suitable treatments are specified in BS 5534 Annexe E.

Note: Where there is a risk that the timber batten moisture content will be greater than 20%, treatments that react with metal fixings should be avoided or alternative metals chosen. If treatment is required, contact the batten manufacturer for advice on fixings to be used.

All roofing battens must be graded to BS 5534. In which clause 4.11 gives a list of permissible species of timber from which battens may be cut and also a table of minimum

Table 2.1 Recommended minimum timber batten sizes (roofing and vertical work)

	Basic minimum size of battens			
	Up to 450 mm span		451 mm – 600 mm span	
	Width mm	Depth mm	Width mm	Depth mm
CLAY AND CONCRETE TILES				
Double-lap	38	25	38	25
Single-lap	38	25	50	25

battens sizes with their permitted tolerances for roofing and vertical cladding applications. Table 2.1 shown below is a simplified version of the BS 5534 original.

For rafter spans greater than 600 mm, BS 5534 Annexe F recommends batten designs including structural calculations for strength and stiffness.

BS 5534 describes permissible characteristics and defects such as knots, fissures, slope of grain, wane, rate of growth, resin pockets, distortion, decay, insect attack, sap-stain and moisture content. Defects must not exceed the limits set down in BS 5534. A knot appearing on opposing batten edges is not permissible if it is greater than 5 mm or 20% of the batten depth.

BS 4978 gives a minimum size for strength grading of timber generally (formerly called stress grading), however, the rules within BS 5534 are specifically designed for tiling battens and the associated loads/requirements. Therefore BS 4978 should not apply to the grading of standard battens.

Concerns

There are continuing concerns about a widespread practice in the timber and roofing trades about using battens which are below the recommended size or have serious defects which pose a safety hazard to the tiler during installation. It may be that they do not provide adequate support to the tiles and other roof loads, and they may not allow secure mechanical fixings to be used without problems.

Over the years, different manufacturers have coloured their battens as a visual aid to prove compliance, but the organisations that supply substandard products are also aware of this and colour their battens even if they do not conform to BS 5534. Do not rely on the colour of a roofing batten to guarantee that it is compliant.

BS 5534 looks to prevent such problems and strongly recommends that the following information is documented and indelibly marked on each batten:

- Name of supplier.
- Origin.
- Graded to BS 5534.
- Basic sizes.
- Type of preservative (if applicable).

The use of graded timber in the building industry for purposes other than slating and tiling is now an accepted procedure. The current BS 5534 reflects the decision that the supply of fully graded timber battens should be enforced. Supplies of factory fully graded battens are now commercially available. Previously, suppliers have largely relied upon the diligence of the craftsmen on site to avoid the use of timber battens which appear to exceed the restrictions regarding size, knots, wane and so on. Claims on disputes on contracts often occur after the roofing work is finished and the contractor who carries out the supply and/or fixing of the battens is normally held responsible for the quality of the battens supplied.

Environmental certification

It is often a contract requirement to use certified wood verifying it is from a legal and sustainable source. The two main certification bodies are the Forest Stewardship Council (FSC) and the Programme for the Endorsement of Forest Certification (PEFC). Central Point of Expertise on Timber Procurement (CPET) requires legal and sustainable timber to be used on most public sector funded projects, and most main contractors have a sustainable procurement policy. Using certified controlled wood provides the most suitable form of evidence. FSC and PEFC certified timber is widely available, but must be requested at the time of order to maintain the chain of custody.

Guidance notes

Significant improvements have been made in BS 5534 in order that battens may be subjected to a level of quality control compatible with other components in the roof assembly:

- Tolerances on basic sizes: width +/–3 mm; depth –0/+3 mm; based on measurement at a reference moisture content of 20%.
- Span is defined as the distance between the face of supports plus half the bearing length at each end support, whichever is the lesser. The minimum end bearing length should be 17.5 mm.
- Battens for spans greater than those given in BS 5534 Table 3 for other slates, tiles and shingles, such as timber shingles and shakes or metal tiles or other proprietary roofing products, should be in accordance with the manufacturer's recommendations.
- Graded battens will be marked and accompanied by the identification information given in BS 5534. Batten sizes for rafter spans greater than 600 mm should be designed by structural calculation in accordance with BS 5534 Annexe F for strength and stiffness. When determining batten sizes, consideration should be given to adequate dimensions for nailing and using commercially economic sizes.
- Where manufacturers give specific batten sizes for fixing only of ridge, hip capping and valleys, these should be followed in those locations. They should not be used for general areas.
- Nails for use with battens and boarding (board sarking) should conform to BS EN 10230–1. For extra protection and in coastal regions, they should be coated by zinc or zinc alloy coating methods specified in BS EN 10230–1.
- Timber containing copper-based preservative can cause corrosion of uncoated mild steel nails in the presence of moisture.

2.7 Roof structure and insulation

Tiles can be put on a wide range of substructures. It would not be possible to illustrate every possible combination of roof structure and tile fixing method. However, the range of examples given here should meet the needs of most new and replacement roofs.

Roof with insulation at ceiling with impermeable underlay

Figure 2.11 shows a traditional roof detail with a ventilated roof space and horizontal ceiling. Ventilation should be provided into the roof space in accordance with Approved Document C of the England building regulations (and its national equivalents in Scotland, Wales and Northern Ireland) and BS 5250.

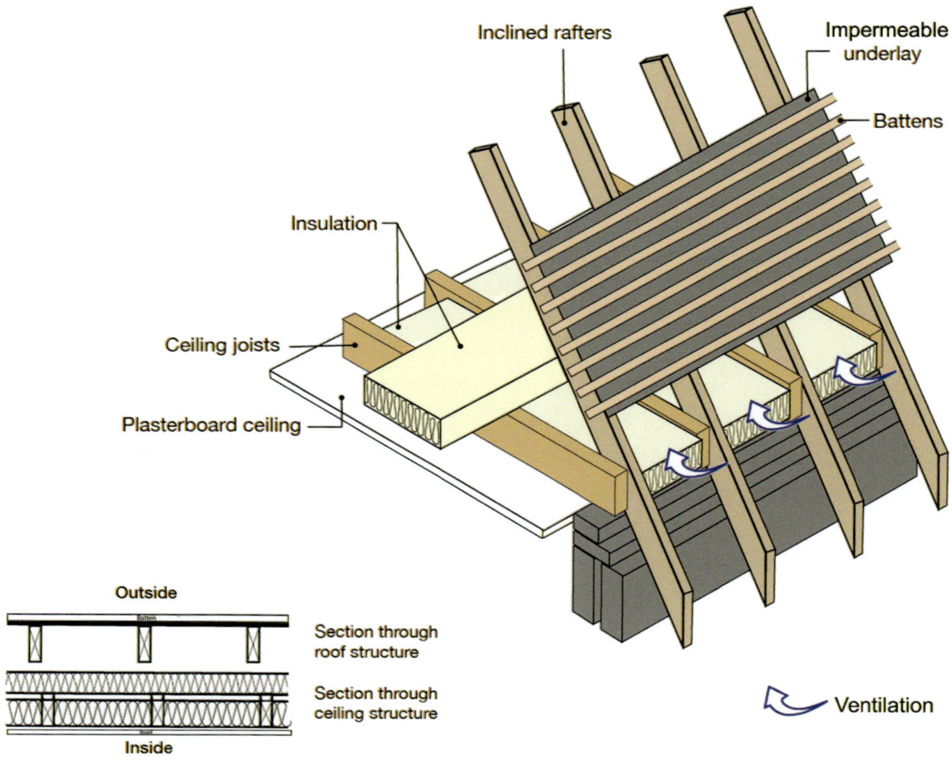

Figure 2.11 Ceiling insulation with impermeable underlay

Roof with insulation at the ceiling with permeable underlay

Figure 2.12 shows roof detail with a horizontal ceiling and with vapour permeable underlay. Ventilation should be provided in accordance with Approved Document C of the England building regulations (and its national equivalents in Scotland, Wales and Northern Ireland) and BS 5250.

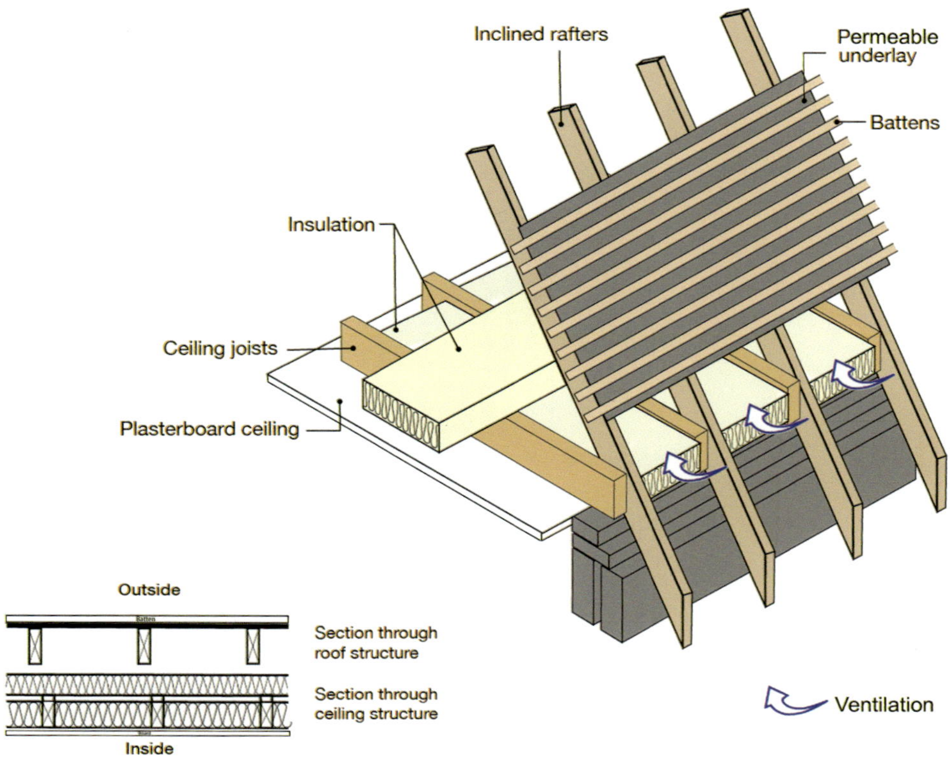

Figure 2.12 Ceiling insulation with permeable underlay

Roof with insulation at rafter with impermeable underlay

Figure 2.13 shows a sloping ceiling and uses a non-vapour permeable underlay and is suitable where there is a room in the roof space. Insulation is fitted between the rafters. It is recommended that a clear ventilated air gap no less than 25 mm must be positively ventilated in accordance with Approved Document C of the England building regulations (and its national equivalents in Scotland, Wales and Northern Ireland) between the underlay and the insulation. To allow for drape, counter battens of 38 mm to 50 mm are recommended.

Roof with insulation at rafter with permeable underlay

Figure 2.14 shows a sloping ceiling and uses a vapour permeable underlay without additional roof space ventilation and is suitable where there is a room in the roof space. Insulation is fitted between the rafters.

A vapour control layer is fitted to the warm side of the insulation in accordance with BS 5250.

An alternate version of this design is that it incorporates an additional layer of insulation below the rafters above the ceiling.

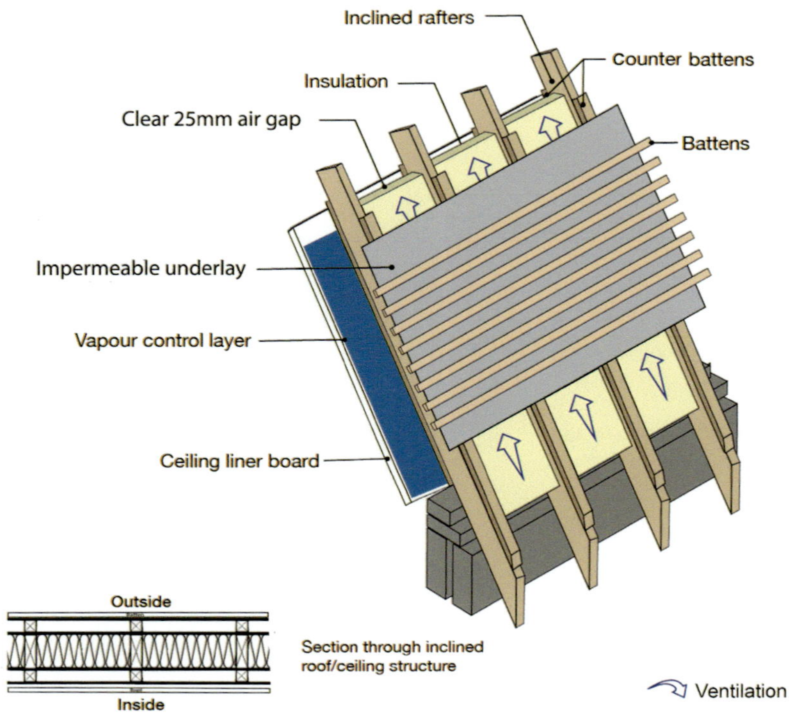

Figure 2.13 Rafter insulation with impermeable underlay

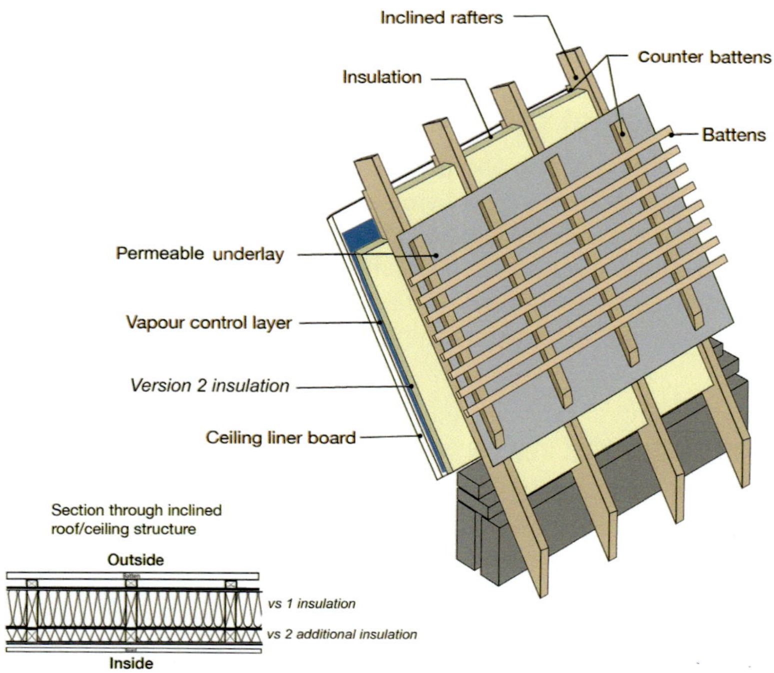

Figure 2.14 Rafter insulation with vapour permeable underlay

Roof with insulated liner system with permeable underlay

Figure 2.15 shows the use of a metal liner tray system fixed to horizontal steel purlins. Rigid insulation is positioned between the upstands of the liner. Vapour permeable underlay is laid over the upstands, and counter battens are secured to the upstands using proprietary fixings as recommended by the liner manufacturer. Tile battens are then nailed to the counter battens in the usual way. In this construction, the metal liner tray acts as the vapour control layer so care must be taken to prevent any unsealed penetrations.

Where a suspended ceiling system is used (not shown), advice should be sought from the liner manufacturer.

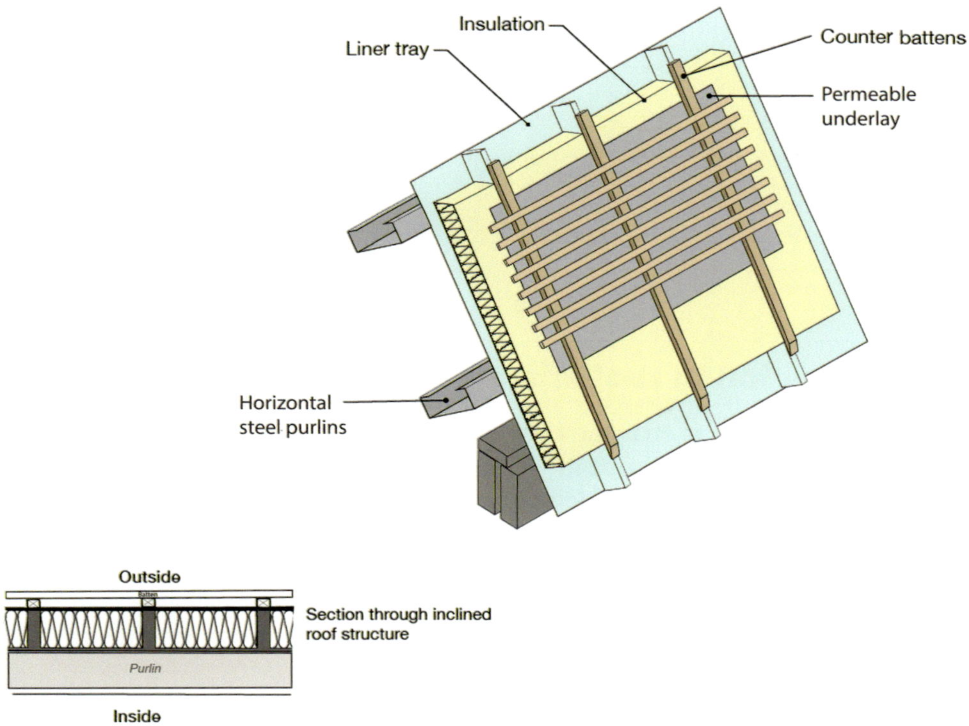

Figure 2.15 Insulated liner tray system

Roof with preformed insulated liner system

Figure 2.16 shows the use of a metal preformed liner system with insulation 'sandwiched' between lower and upper metal panels. The system is fixed to horizontal steel purlins. Tile battens are secured to the counter battens.

BS 5250 states that there is no risk of surface condensation with such panels, nor would interstitial condensation occur, provided the vapour resistance of the inner surface layer is equal to or greater than that of the outer layer and all joints between the panels are sealed.

Where a suspended ceiling system is used (not shown), advice should be sought from the liner manufacturer.

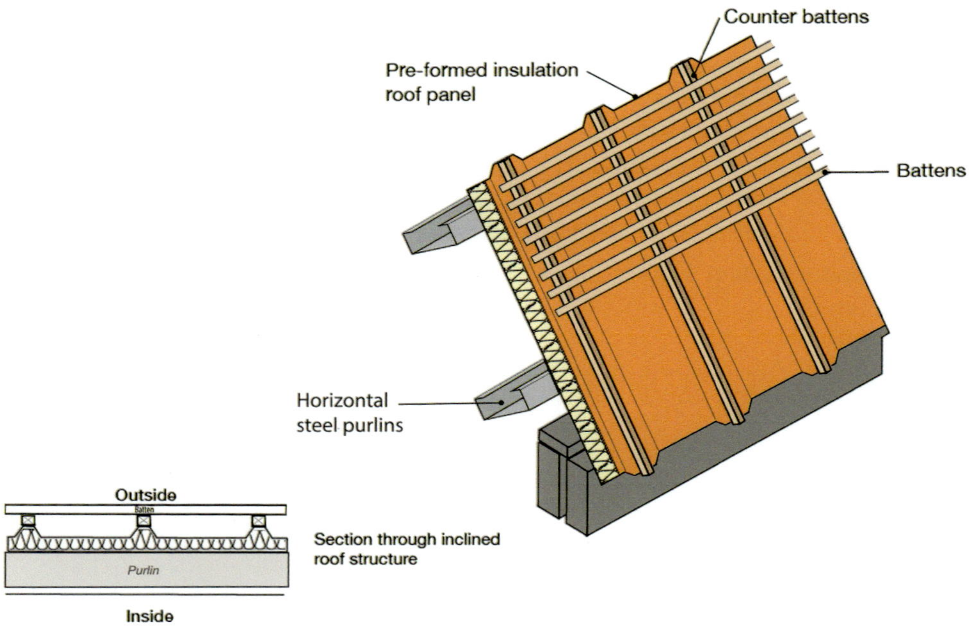

Figure 2.16 Preformed insulated liner tray system

2.8 Condensation in roof spaces

Moisture risk in buildings has changed considerably over the past few years as new methods of construction and occupants' lifestyles have changed. Through activities including breathing, washing and cooking the average home produces roughly 10 litres of moisture every day. The moisture created can cause an issue when the external temperature is low. This is because we close windows to keep the warmth in, turn the heating up and dry our clothes inside. When the water vapour has no way to escape the building interior, condensation can form on all cold surfaces, which, if not properly controlled, can result in mould growth and damage to the building fabric.

Before 2021, BS 5250 focused mainly on condensation, but in 2021 it was updated to include other moisture problems such as excessive humidity, rising damp, rain penetration and roof leaks.

It also recognises that, in both new and existing buildings, the gap between the 'As Designed Theoretical' (ADT) and the way it performs when it is built and in use 'As Built In Service' (ABIS) may vary significantly.

Why does condensation occur in roof spaces?

The simple science behind why condensation occurs is that warm air can hold more moisture than cold air.

A warm living space with a temperature of 20°C has the capacity to hold 20 g of water per cubic metre of air, but a cold roof space at 10°C can only hold 8 g. So, when the warm air rises up into the roof, 12 g of water per cubic metre of air needs to go somewhere. When a surface is cooler than the dew point temperature (when the air is fully saturated with water vapour), condensation will occur as droplets, typically on the underside of the roof underlay.

Condensation occurs in roof spaces when there is too much warm air circulating upwards from living spaces and/or when there is inadequate roofing ventilation. Older buildings tend to have a degree of natural ventilation due to less accurate building methods, therefore offering more opportunities for moisture to escape. As newer builds have a much better thermal performance in a bid to reduce energy consumption, roof spaces tend to be colder, and the risk of condensation is increased.

This is the reason why building regulations state that 'roofs should be designed and constructed so that their structural and thermal performance are not adversely affected by interstitial condensation'.

When does condensation become a problem?

In a correctly constructed roof, condensation may occur on a temporary basis due to extremely cold weather, when people are keen to keep warmth in and cold out by limiting air exchange in the home. This is a common occurrence, and this condensation usually dissipates from the underlay within a few days with no harm done. However, when this condensation builds up it can make other roof elements such as rafters and insulation wet, causing water staining on the ceiling or structural damage to the roof.

What are the solutions?

Condensation can be controlled within the roof structure through a combination of limiting airflow from within the living spaces into the roof space, providing controlled airflow from outside and the use of suitable roof underlays. Water vapour can be prevented from reaching the roof space as much as possible by ensuring that there is adequate ventilation and by reducing the amount of moisture generated in a building.

England Building Regulation Approved Document C-2 states that 'the floors, walls and roof of the building shall adequately protect the building and the people who use the building from harmful effects caused by:

- Ground moisture.
- Precipitation including wind-driven spray.
- Interstitial and surface condensation.
- Spillages of water from, or associated with, sanitary fittings or fixed appliances'.

The Approved Document refers to BS 5250 as a means of achieving compliance with the building regulation requirements. Other methods for controlling condensation can be considered, for example through the use of vapour and air-permeable underlays, however, the product must have third party certification by a UKAS accredited body and the design and application recommendations provided in the product certificate must be followed.

BS 5250 advises that 'occupants often fail to use a building as intended and designers are advised to err on the side of caution and adopt robust fail-safe solutions'. Therefore, a condensation risk analysis should be done at the start of any pitched roofing project, identifying appropriate measures to control the release of moisture from the building and how to adequately protect the roof in accordance with the recommendations contained in BS 5250.

How can roof space be ventilated?

There are many roof ventilation products on the market, including eave vents, concealed tiles, ridge and top abutment ventilators. Roof geometry and pitch can dictate where these are best placed, but cross-flow ventilation is usually improved by ventilators installed just above the horizontal insulation at each side of the roof, typically at the eaves. Further high-level ventilators will help draw air in through lower ventilators creating a circulation path for airflow.

How does underlay affect condensation?

In a roof space, condensation typically appears on the underside of roof underlay as it is a cold surface. Temporary condensation is expected when it occurs for a few days in cold weather, or when the building fabric of a new build home is drying out during the first 12 months following completion. However, this build up of condensation can be avoided through installing passive ventilation systems and choosing an appropriate underlay, such as:

- non-breathable high resistance underlay (HR)
- vapour permeable low resistance underlay (LR)
- air and vapour permeable low resistance underlay (LR)

Refer to Chapter 4.1 *Underlay* for more detailed information.

Correct use of underlays

All types of pitched roofing underlay need to be installed correctly in order to perform as designed, and it is essential to consult the manufacturer's installation guidance. Not all underlays require roof space ventilation and the manufacturer can provide evidence of underlay suitability in the given location.

A common mistake when installing an underlay is that laps are not secured using battens or tape (which may be necessary to comply with BS 5534).

Underlays of any type are not necessarily designed to substitute for adequate condensation control measures and appropriate levels of roof ventilation.

2.9 Ventilation in cold and warm roofs

Cold and warm roofs

- A cold roof is historically the most common pitch roof construction in the UK where insulation is laid at ceiling level between or on the joists, so no heat from the building below gets into the loft space.
- A warm roof is becoming increasingly popular due to habitable rooms being built into the loft space. Insulation is placed between and/or above the roof joists allowing the whole building to retain heat.

Well-sealed ceilings (continuous ceilings)

A well-sealed ceiling is constructed as a 'normal ceiling' but with additional sealing to limit the passage of air (containing heat and moisture) through its structure.

Its design avoids construction holes and gaps at wall and ceiling junctions, gaps around pipe and cable penetrations, and air leakage through loft hatches, downlights and such like.

It is essential that cables, lights or pipes that pass through the ceiling be sealed, otherwise, any heat and moisture will have a passageway into the roof voids. All penetrations should be permanently sealed with suitable proprietary products. The ceiling is sealed to the external walls to again limit any leakage through gaps/cracks.

Downlights should comply with BS EN 60529 and be rated IP60 to IP65 (depending on room use) or incorporate an appropriate sealed hood or box which meets the following test criteria:

- The total leakage through downlighters should not exceed 0.06 m^3/hr/m^2 of ceiling at 2.0 Pa.
- The heads of all cavities in the external walls, party walls and partition walls should be sealed to prevent the transfer of warm moist air into the loft.

No access door or hatch should be located in rooms where large amounts of moisture are produced, including kitchens or bathrooms. The air leakage rate through an access hatch, including its frame, when tested to BS EN 13141–1 should be less than 1.0 m^3/hr at a pressure difference of 2.0 Pascals (Pa). It can be assumed that 'push-up' wooden hatch covers in a frame, constructed in situ, with continuous compressible seals, will meet this criterion provided the weight of the door is at least 5.5 kg.

Hatch covers should either be heavy enough to compress a seal or be clamped, with a closed cell compressible seal, or an 'o-ring' between it and the frame. Drop-down hatch covers are more difficult to seal and it is recommended that propriety units with a supplied hatch cover in a frame are used. Manufacturers can provide third-party evidence that the leakage criterion is met.

If there is any doubt, assume that the ceilings are not well-sealed and design and construct the roof accordingly. BS 9250 gives further practical advice on this.

Ventilation requirements for roof spaces

Roof spaces should be ventilated in accordance with BS 5250, with 3 mm, 5 mm, 7 mm, 10 mm, and 25 mm continuous ventilation products giving 3000, 5000, 7000, 10000 and 25000 mm^2/m respectively. Non-continuous products such as tile vents may be used spaced apart so as to achieve the equivalent ventilation requirement.

Fully boarded roofs

On pitched roofs, fully boarded (lined with ply), that are classified as air impermeable, all underlays should be treated as impermeable and the roof space below should be ventilated in accordance with the high resistance underlay requirements described below.

On fully boarded roofs that are classified as air-open, such as traditional Scottish sarking boards with air gaps, these are classed as vapour permeable in line with BS5250, therefore the underlay can be vapour permeable and the roof space ventilation can be designed in accordance with low resistance underlay requirements.

Cold roof construction designs

Vapour impermeable HR (non-breather) underlay

HR underlays have a high resistance to the passage of water vapour (more than 0.25 MNs/g). Also commonly referred to as 'non-breather' or 'impermeable' underlays. The following diagrams show its use in typical roof constructs.

Figure 2.17 shows that, if a building is less than 10 metres wide and the roof pitch is less than 35° then 10,000 mm²/m, eaves ventilation is required.

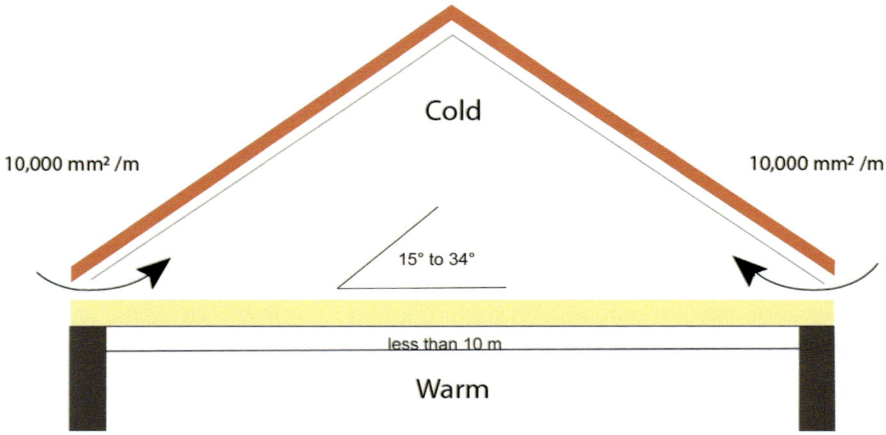

Figure 2.17 Cold roof with HR underlay – Pitches of 35° or less with a span of 10 m or less

Figure 2.18 shows that, if a building is 10 metres wide or more or the roof pitch is 35° or above, then 10,000 mm²/m eaves ventilation is required together with 5,000 mm²/m ridge ventilation.

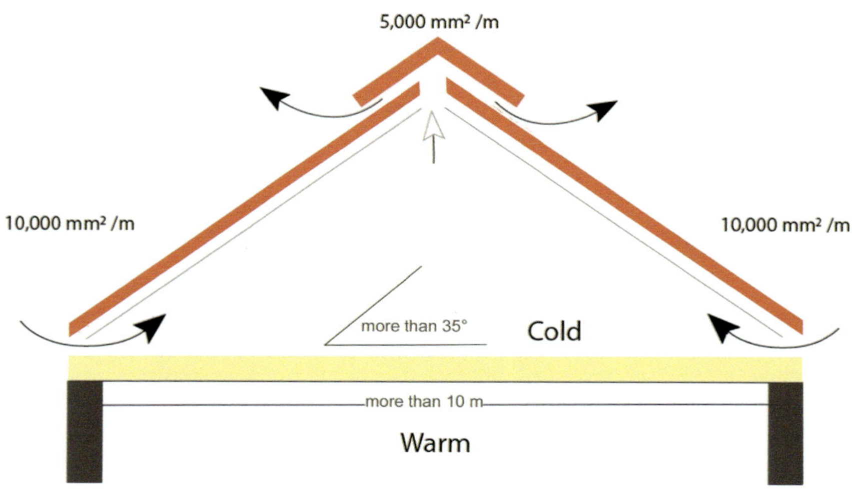

Figure 2.18 Cold roof with HR underlay – Pitches over 35° with a span greater than 10 m

Figure 2.19 shows that, if a roof pitch is 15° or less, then 25,000 mm²/m eaves ventilation is required.

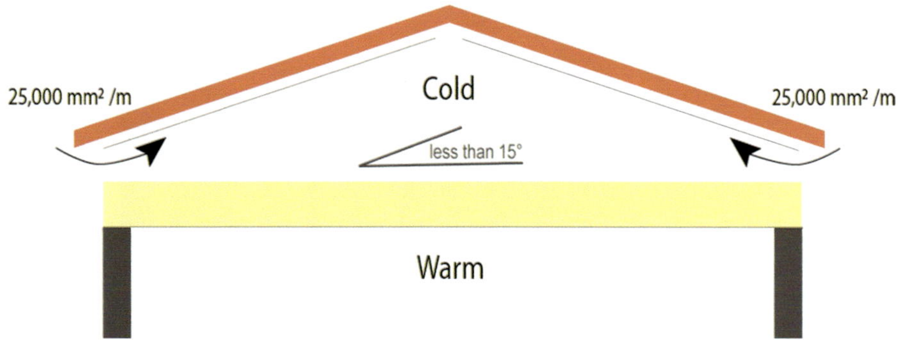

Figure 2.19 Cold roof with HR underlay – Pitches of 15° or less

Figure 2.20 shows a monopitch or lean-to roof, with a pitch of 15° or more, where 10,000 mm²/m eaves ventilation is required together with 5,000 mm²/m ridge or top edge ventilation.

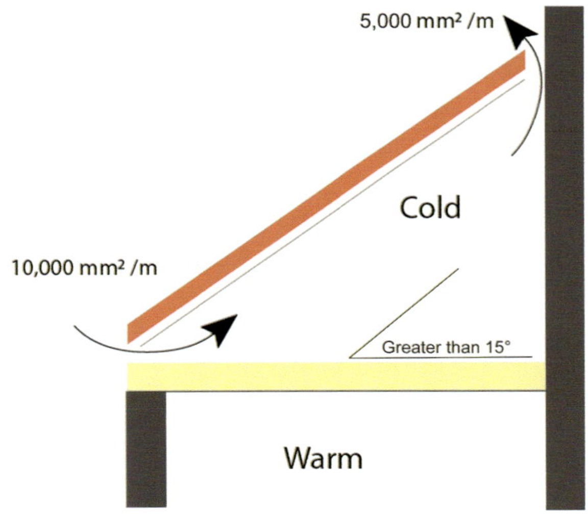

Figure 2.20 Cold roof with HR underlay – Monopitch greater than 15°

Figure 2.21 below shows a monopitch or lean-to roof with a pitch of less than 15°, when 25,000 mm²/m eaves ventilation is required together with 5,000 mm²/m ridge or top edge ventilation.

Vapour permeable LR (breather) underlays

LR underlays have a low resistance to the passage of moisture vapour (less than 0.25 MNs/g) which allows the transfer of water vapour. Commonly referred to as 'breather' or 'permeable'

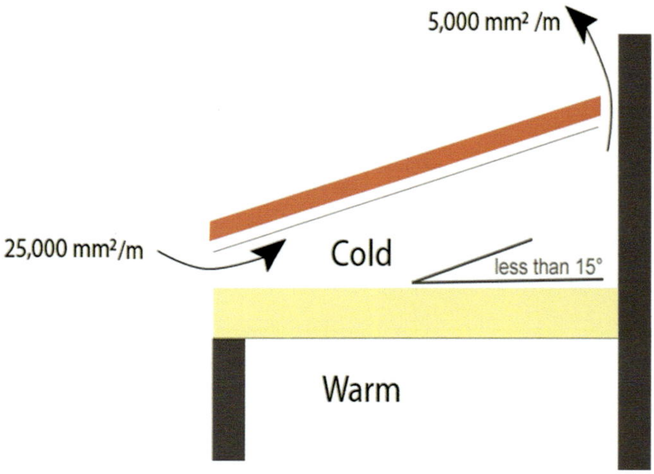

Figure 2.21 Cold roof with HR underlay – Monopitch less than 15°

underlays, they can contribute to reducing condensation risk. The following diagrams show its use in typical roof constructs.

In these types of roof construction, where the external covering of the roof consists of close-fitting tiles or slates which are considered relatively airtight (such as fibre cement or rubber) and would resist the free movement of water vapour to outside air, the roof should be fully ventilated in the same manner as with an HR underlay. Natural clay and concrete tiles are generally considered air-open and suitable for this type of installation.

Figure 2.22 shows that, if a building has a well-sealed ceiling, then 3,000 mm^2/m eaves ventilation is required. If a building has a normal ceiling, then 7,000 mm^2/m eaves ventilation is required. In practice, a commercially available 10,000 mm^2/m eaves ventilation system would normally be used. It should be noted that:

- UKAS accredited vapour permeable membranes are available that can be used without eaves or ridge ventilation.

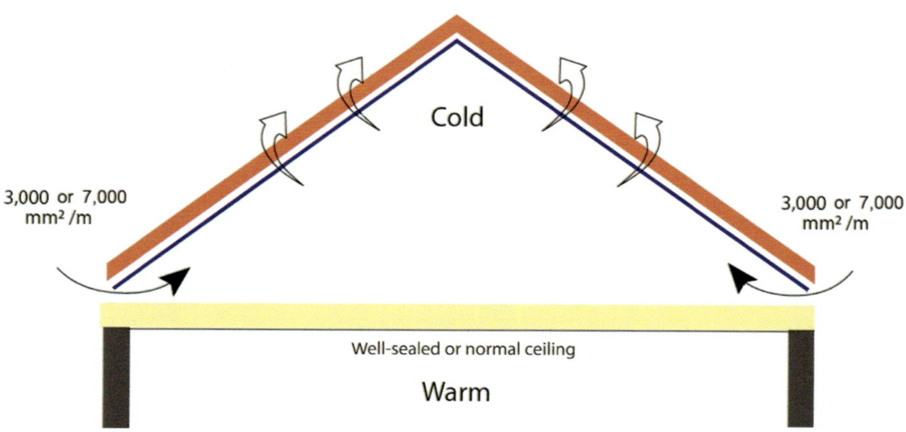

Figure 2.22 Cold roof with LR underlay – All pitches with any roof spans

Figure 2.23 below shows as an alternative, when the building has only a well-sealed ceiling, 5,000 mm2/m ridge (or other high level) ventilation can be used. In this design, it should be noted that:

- High-level ventilation is a mandatory requirement by NHBC.

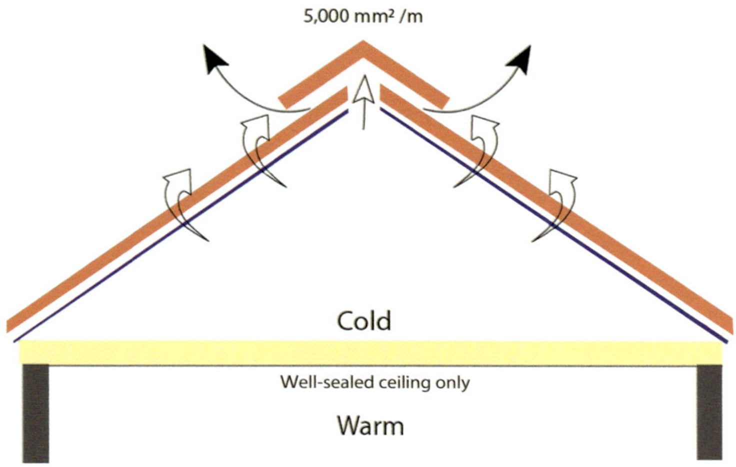

Figure 2.23 Cold roof with LR underlay – well-sealed ceilings only alternative

Figure 2.24 shows that, in buildings larger than dwellings, in addition to the 5,000 mm²/m ridge ventilation, eaves ventilation is also required. With a well-sealed ceiling, eaves ventilation of 5,000 mm²/m is needed. With a normal ceiling, 10,000 mm²/m of eaves ventilation is needed.

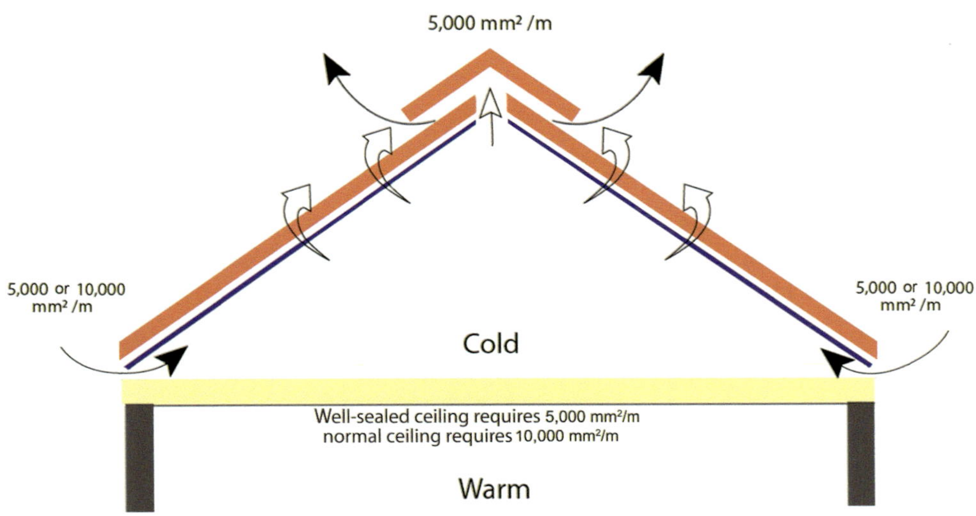

Figure 2.24 Cold roof with LR underlay – larger than dwelling buildings

Warm roof construction designs

All warm roof constructions require a continuous air and vapour control layer (AVCL) to be installed on the warm side of the insulation. All laps, joins and penetrations should be fully sealed.

It is important to note that if a well-sealed ceiling is not provided and/or the integrity of the AVCL cannot be maintained then, regardless of the underlay type used, a ventilated air space should be provided.

Vapour impermeable HR (non-breather) underlay

Figure 2.25 shows a well-sealed ceiling with an AVCL. There should be a 50 mm gap between the underlay and insulation, to ensure a minimum 25 mm gap at the centre of the underlay drape, and 25,000 mm²/m eaves ventilation and 5,000 mm²/m ridge ventilation is required.

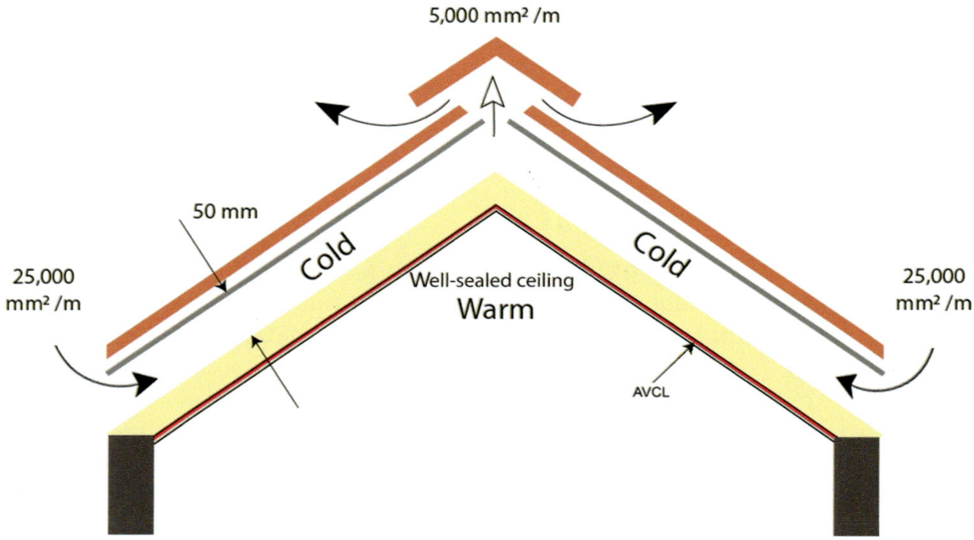

Figure 2.25 Warm roof with HR underlay

Figure 2.26 shows that, where there are obstructions to the airflow, such as at firewalls or roof windows, additional ventilation of 5,000 mm²/m below and 25,000 mm²/m above the obstructions is required.

Figure 2.27 shows that, where the insulation only partially follows the roof slope, such as with room-in-roof construction, then 25,000 mm²/m eaves ventilation and 5,000 mm²/m ridge ventilation are required.

Vapour permeable LR (breather) underlays

Figure 2.28 shows the installation (and maintenance) of a well-sealed ceiling with a vapour-permeable underlay and an efficient AVCL. In this case, ventilation is not required. If it is not possible to install an efficient AVCL then a ventilated roof void should be installed as shown

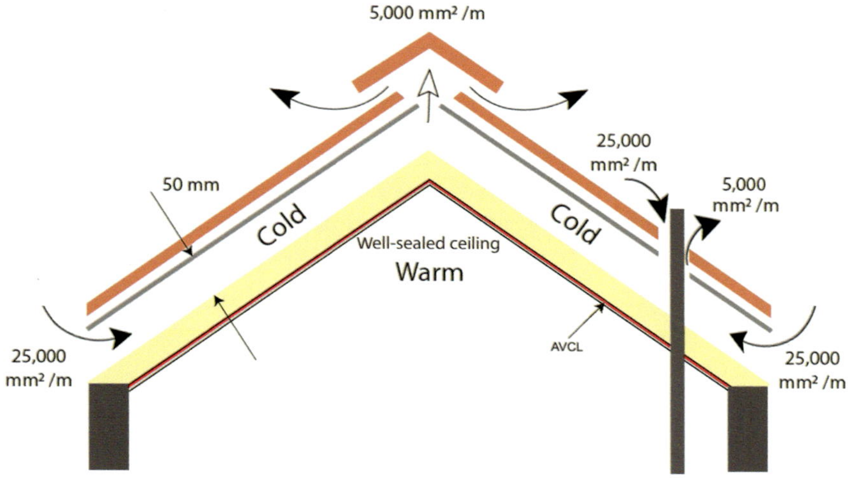

Figure 2.26 Warm roof with HR underlay – with obstruction to airflow

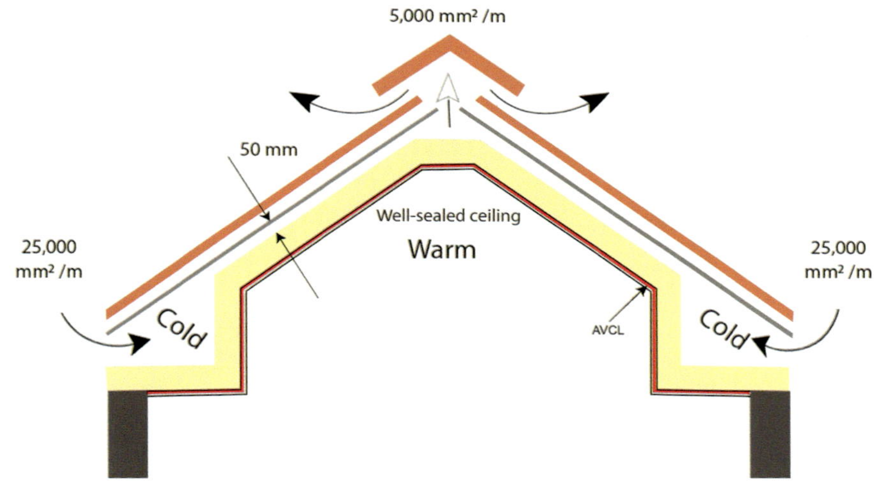

Figure 2.27 Warm roof with HR underlay – with irregular shape ceiling

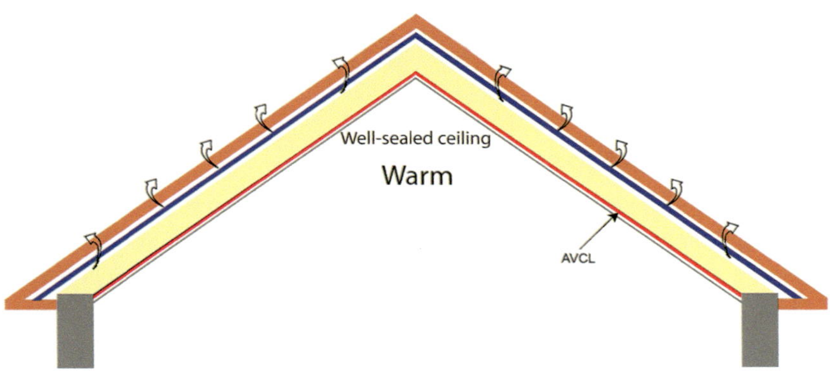

Figure 2.28 Warm roof with LR underlay- well-sealed ceiling and effective AVCL

in Figure 2.25. The vapour permeable underlay may be laid fully supported on the insulation or draped unsupported.

It should again be noted that, with this type of warm roof construction, as was the case with cold roof and LR underlay construction, where the external covering of the roof consists of close-fitting tiles or slates which are considered relatively airtight (such as fibre cement or rubber) and would resist the free movement of water vapour to outside air, the roof should be fully ventilated in the same manner as with an HR underlay. Natural clay and concrete tiles are generally considered air-open and therefore suitable for this type of installation.

Figure 2.29 shows that, where the building has a normal ceiling rather than a well-sealed ceiling, then 25,000 mm²/m eaves ventilation and 5,000 mm²/m ridge ventilation are required, unless the underlay has a relevant UKAS-accredited non-ventilated certificate.

This certification allows LR (breathable) underlay to be used without ventilation (as shown in Figure 2.28) provided a number of conditions are met. Principal among the requirements are:

- the products can only be used in dwellings (large buildings, schools, factories and shops are not permitted)
- a convection-tight loft space shall be achieved (it is recognised that the high standard required to achieve and maintain a 'convection-tight' loft space would be extremely difficult to reach)

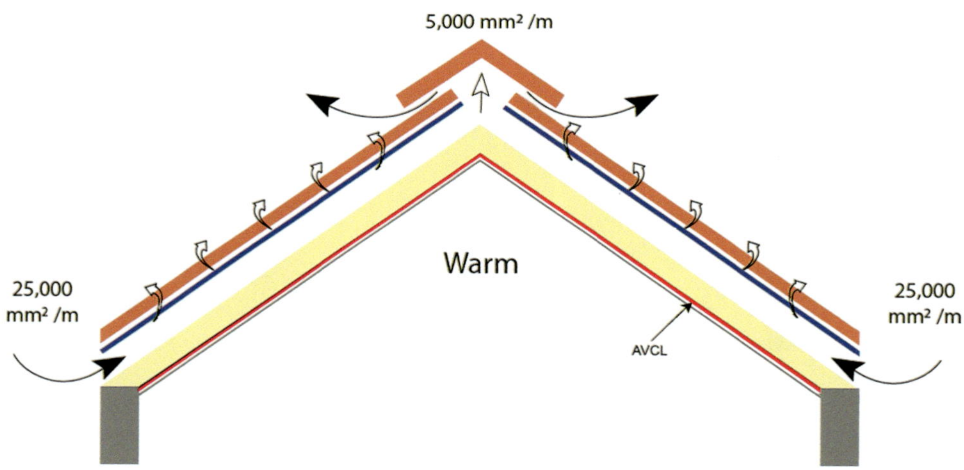

Figure 2.29 Warm roof with LR underlay – normal ceiling and AVCL

Energy efficiency and loss control

In conventionally ventilated pitched roof constructions, energy loss by ventilation can account for up to 25% of the total heat loss through the roof. The BBA has stated that a non-ventilated roof system using a vapour-permeable underlay will significantly reduce this heat loss.

Around 50% of the heat lost through the ceiling (and 80% of the moisture transferred to the roof space) is through air leaking through holes and cracks in the ceiling. The balance of the heat loss and moisture transfer is by conduction and diffusion through the fabric of the ceiling.

As described earlier in this ventilation chapter, BS 5250 allows for the concept of a well-sealed ceiling to be used to reduce the moisture transfer into the roof space, and this type of construction can also reduce heat (energy) loss from the building into the roof space. BS 5250 defines this type of ceiling as 'The design avoids constructional gaps, especially at the wall ceiling junction with dry lining construction, and holes in the ceiling'. Therefore, the well-sealed ceiling simultaneously contributes to the air tightness (energy efficiency) of the building.

2.10 Valley design

The pitch at valleys is, on average, around 5 degrees lower than the roof pitch, and as such is an area which is potentially vulnerable to water ingress. The purpose of this section is to provide appropriate guidance on the installation design of valleys to ensure that the risk of failure is reduced to a minimum.

Design and testing background

The information in this chapter is based on the results of an industry research project, which was undertaken at Brunel University in 1990, looking at the effects of rainwater on inclined valleys in order to determine the minimum open channel widths. Both the conclusions and recommendations of the testing were subsequently used in the drafting of the design recommendations for roof drainage in BS 5534, BS 8000–6, the LSTA *Rolled Lead Sheet – The Complete Manual* and NFRC technical bulletin TB28.

Criteria for testing

When designing pitched roof valleys, it is vital to take into consideration a number of factors to make sure that adequate drainage is provided for the roof areas to be discharged of rainwater. The summary of the investigations performed on test rigs constructed at Brunel University was as follows:

- Two roof rigs were constructed with variable intersecting pitched roof slopes to simulate the lower 2.5 m valley section with the dimensions of:
  - a) 5 m x 5 m roof area on plan
  - b) 10 m x 10 m roof area on plan.
- A design rainfall rate of 225 mm per hour on plan was used, as this was believed to be the worst-case scenario that could occur for 2 minutes once in 50 years. See BS EN 12056–3 for more detail.
- Rafter pitches from 12.5° to 35° were investigated.
- The effect of flow from profiled tiles and flat tiles was assessed.
- Valley widths from 100 mm to 300 mm were investigated, and the effect of different valley product materials was assessed.

Conclusions and recommendations

The research found that the ability of the open valley channel to collect water and drain it away safely will depend on the volume of rainfall, the true pitch of the valley, and the area of the roof served by the inclined valley. In particular:

- The failure was assumed to be the same for bedded or dry (non-mortared) valleys.
- The different valley product materials had a negligible effect on the valley flow rates.
- Profiled tiles or flat tiles had a negligible effect on the valley flow rates.
- The design data will apply where the two roof pitches, adjacent to the valley, have a difference equal to or less than 5°. The lower pitch should be used for assessing the valley design data. Where the difference between roof pitches adjacent to the valley is greater than 5°, it will be necessary to seek specific design advice from the manufacturer.
- The design data only applies to rafter lengths equal or less than 5 m on plan, and greater than 5 m plus equal or less than 10 m on plan.
- The design data only applies to the total two roof areas discharging into the valley which have a difference in plan area not greater than 10%. In cases where the two roof areas discharging into the valley are greater than 10%, advice must be obtained from the manufacturer.
- Where vertical walls or projections drain into a roof, it is necessary to add 50% of the vertical wall area onto the plan area of the roof onto which it discharges.
- The design data applies to open valleys and does not apply to soaker closed or mitred valleys, secret gutters or box gutters.

Valley designs

The point/joint at which two different roof aspects meet forms a drainage channel which is referred to as the valley. It can be considered one of the most vulnerable areas of a roof, as it needs to carry away heavy rainfall without leakage or blockage. There are two main types of valleys.

- Closed – the valley is protected by, and covered by, the roofing tiles.

Standard tiles can be cut to form swept, mitred or laced covering. Purpose-made valley tiles can be used to form the water runoff. Additional underlay or metal (or similar) soakers may also be required to gain the weathertight integrity. In all cases, tiles should be cut neatly to form a smooth junction, preferably cutting from tile-and-a-half tiles to avoid small-cut tiles.

- Open – the valley is not covered by the tiles.

The valley trough is formed from lead on site, or preformed offsite in non-ferrous metal or glass reinforced polyester (GRP). A common alternative to the traditional wide open profile shape is the inclusion of a central upstand, with channels at each side of the upstand to direct water towards the gutter without leaking over its edges. This design is beneficial when used between two roof slopes of unequal pitches and also, as the tiling is finished close to each side of the upstand, the adjacent tiles are much closer together thereby giving the appearance of a closer-cut roof.

Chapter 6.5 *Valleys and gutters* has more information on valley installations.

Valley gutter designs

The following Table 2.2 shows the recommended minimum widths of valley gutters for different roof pitches at different rates of rainfall. There are three categories of design rate of rainfall to be considered as summarised in BS EN 12056–3:

Table 2.2 Recommended minimum widths of valley gutters for different roof pitches

Recommended minimum widths of valley gutters for different roof pitches						
Roof pitch (degrees)	**Design rainfall rate**					
	225 mm/h		**150 mm/h**		**75 mm/h**	
	Equal or less than 25 m² on plan	**Greater than 25 m² but equal or less than 100 m² on plan**	**Equal or less than 25 m² on plan**	**Greater than 25 m² but equal or less than 100 m² on plan**	**Equal or less than 25 m² on plan**	**Greater than 25 m² but equal or less than 100 m² on plan**
12.5–17	150	250	125	200	125	150
17.5–22	125	200	125	150	100	125
22.5–29	100	150	100	125	100	100
30–34	100	125	100	100	100	100
35 +	100	100	100	100	100	100

- V0.022L/sec per m² (225 mm/h).
- V0.015L/sec per m² (150 mm/h).
- 0.010L/sec per m² (75 mm/h).

The designer should choose the rate of rainfall that, at the chosen location, has a period equal to or greater than the required return period. When in doubt, the worst-case design rate of 225 mm/h or 0.022 L/sec per m² should be used.

All valley gutter widths assume that adjacent roof slopes are at 90° to each other on plan. Where this is not the case, additional calculations may be required to determine the valley width.

Figure 2.30 and Figure 2.31 are given to identify the width of the valley gutters. The dimension A is measured as a horizontal distance between the tiles in millimetres.

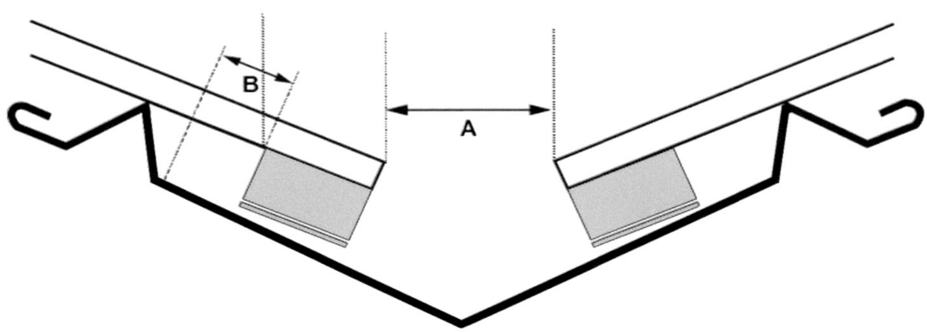

Figure 2.30 Valley with bedded tiles

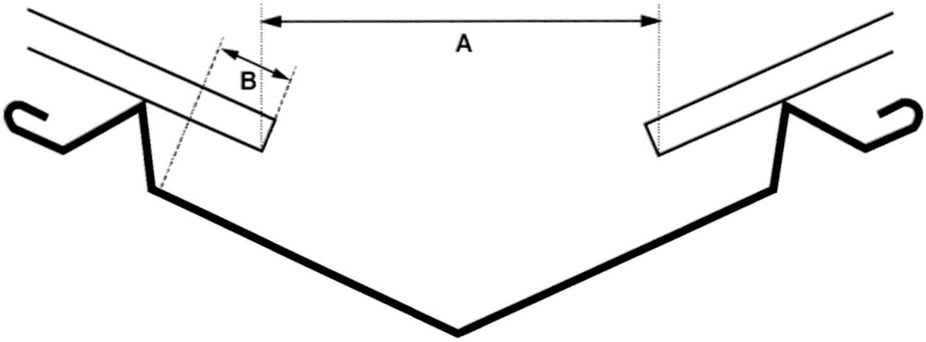

Figure 2.31 Valley with unbedded tiles

Table 2.3 Widths of lead for bedded and unbedded lead-lined valley gutters

Widths of lead for bedded and unbedded lead-lined valley gutters		
Valley width in mm [dimension A in gs 2.19 & 2.20]	**Widths of lead for bedded valley (mm)**	**Width of lead for unbedded valley (mm)**
100	500	400
125	525	425
150	550	450
200	600	500
250	650	550

Table 2.4 Gutter trough angles in degrees

Gutter trough angles in degrees		
Roof pitch	**Valley pitch**	**Trough angle**
12.5	8.9	162.6
15	10.7	159.0
20	14.4	152.0
25	18.3	145.0
30	22.2	138.6
35	26.3	133.2
40	30.7	127.8
45	35.3	120.0
50	40.1	114.4
55	45.3	109.2
60	50.8	10.4.4
65	56.6	100.2
70	62.8	96.8
75	69.2	93.8
80	76	91.8
85	82.9	90.4
90	90	90.0

CHAPTER 3

Types of roof tiles

There are many types of roof tiles used in the UK in terms of the type of material the tiles are made from, and the shape and size of the tiles represent both traditionally shaped single-lap pantiles, double Romans and flat slates to plain tiles and modern designs that replicate older shapes in a more cost-effective way.

The most popular by far are made from either concrete or clay. Each has its own advantages and very often the choice falls to planning consent and/or aesthetics.

3.1 History and development

The evolution of roof tiles as products to cover pitched roofs can be traced back to ancient Egypt, where shaped and curved clay tiles were made by hand and cured by the heat of the sun and laid on roofs, fitted overlapping.

The Chinese also developed coloured glazed tiles with wide, curved under tiles and semi-cylindrical over tiles.

This simple overlapping principle was later developed by the Romans as the 'tegula' (under tile) and 'imbrex' (rain shower) over or capping tile. The Moors developed this design, now called 'Spanish' tiles, using semicircular tapered profiled tiles which allowed alternate rows to be reversed with the convex row seated over the joint of the concave tiles. Later developments with machine-made clay single-lap tiles saw the introduction of sophisticated interlocking tiles with anti-capillary devices at the top and side-laps.

In Britain and Northern Europe during the 13th Century, plain tiles comprising simple, small, rectangular thin strips of flat or cambered clay started to be used as an alternative to thatch and wooden shingles due to their incombustibility. These were laid on the roof in a cross-bonded double-lap pattern.

Later, in the 17th and 18th Centuries, single-lap clay pantiles were imported from Holland, which led to the production of English pantiles around the river Humber and in Bridgwater, Somerset, where estuarine clay deposits could be found.

In the early 20th Century, concrete was introduced in the UK as a material for making roof tiles, and this saw the birth of a whole manufacturing industry. Initially, products were based on handmade double-lap plain tiles, then later on also with machine-made, large format, interlocking, single-lap tiles.

3.2 Concrete

Concrete tiles are one of the most common roofing materials used in the UK and were originally introduced to the UK in the 1920s, but were not particularly popular. However,

DOI: 10.1201/9781003196990-3

after World War 2, an extensive rehousing programme was started, causing high demand for roof tiles. It was the widespread availability of concrete roof tiles and their cheaper cost, that enabled them to become popular as the government pushed forward its rapid house-building agenda.

Concrete tiles are made with a mixture of sand, cement and water, that is shaped and cured into a durable tile. The finished surface of concrete can be either smooth or granular, and a variety of colours can be obtained by the use of pigments or surface treatments for a more unique look.

Concrete tiles perform well in the changeable UK climate and the combination of relatively simple manufacturing process, longevity and affordability has ensured that they have retained their popularity over the years.

Advantages of concrete roof tiles

- More cost-effective roof covering.
- Available in a large range of colours.
- Long warranty periods (in excess of 50 years is common).
- A uniform appearance suited to modern homes.
- Low maintenance.

Disadvantages of concrete roof tiles

- Heavier than equivalent clay tiles, making the cost of the roof structure higher.
- Colour fades over time.

3.3 Clay

Clay is one of the oldest materials used in roofing. Some of the earliest examples of clay roof tiles found by archaeologists were in China and date back as far as 10,000 BC. Nowadays, roof tiles made of clay are used extensively the world over.

Clay is a natural material made up largely of inorganic clay minerals (kaolinite, illite, montmorillonite, etc.) and quartz (sand), along with smaller quantities of other components such as iron oxides, carbonates, pyrite, organic material, etc. It is found in abundant quantities across the South of England, the Midlands and Eastern counties of the UK as a result of geological formations laid down many millions of years ago. There are many different seams of clay (Etruria Marl, Lower Oxford, Weald, London) to name but a few. Each has its own mineralogy and chemical composition and subsequently fires differently in the kiln. Clay deposits inland are better for making small format plain tiles, whilst estuarine clays provide a better consistency for larger single-lap pantiles and double Roman shapes. All manufacturers can offer a range of colours post firing and will often blend two or more different clays to help achieve different tile colours, better tensile strength and higher frost resistance.

Clay tiles can be moulded into different shapes and forms before being fired in a kiln for use on roofs, or as an exterior wall cladding to add to the aesthetics of the property.

Advantages of clay roof tiles

- A natural, rustic appearance that will improve over time with colour that will not fade.
- Range of sizes to suit all roof types and details.
- Range of price, from low-cost machine made to the more expensive handmade.
- Long lifespan with most manufacturers giving at least 30 years warranty.
- Low maintenance.
- Perfect for traditional buildings, renovations and architectural new builds.

Disadvantages of clay roof tiles

- Costlier than concrete tiles (although the cost gap is reducing).
- Colour tone can vary over batches of production and careful mixing is needed on site.

3.4 Manufacturing methods

Machine-made concrete tiles

Concrete tiles should be manufactured to the requirements of:

- BS EN 490 *Concrete roofing tiles and fittings for roof covering and wall cladding – Product specifications*.

There is also a separate standard which contains details of the various tests to be undertaken to demonstrate satisfactory criteria for dimensions, flatness, transverse strength, water impermeability, freeze-thaw and nib support.

- BS EN 491 *Concrete roofing tiles and fittings – Test Methods*.

The process of producing concrete tiles begins with mixing a large batch of either raw or pigmented concrete in an industrial mixing vat. This concrete is then transported by conveyor belts to a production line. On the production line, the concrete mix falls into an extruder, which applies it at pressure onto a continuous row of metal pallets/moulds each shaped to the underside of the tile. As the pallets/moulds move along the production line at high speed, they pass underneath a profiled unit which forms the top of the tile.

Concrete tiles are cured at a much lower temperature than clay tiles as they do not require firing in a kiln. This process takes place in a curing chamber at a temperature of around 37°C. Moisture may also be added to the chamber air to optimise the humidity of the curing environment.

Once cured, depending on the tile surface finish, the tiles are either surface coated with paint or with coloured sand granules. The tiles are then moved to a drying cabinet, which allows sufficient time for the surface coating to harden. Once the tiles are dry, they are then moved to a packing area to be inspected, packed and stored prior to distribution.

Although most concrete tiles are machine-made, some manufacturers produce handmade fittings to complement main roof tiles. These can include items such as purpose-made hip and valley tiles, which are custom-made to suit the pitch and angles of the roof.

These handmade fittings are formed starting with a semi-dry concrete mix which is placed on a jig or mould and shaped to angle and dimensions to suit the design requirements of the roof tile or fitting. They are then finished with either a granular surface or pigmented coating, and placed in a curing chamber to cure before packaging and distribution to the customer.

Machine-made clay tiles

Clay tiles should be manufactured to the requirements of BS EN 1304 *Clay roofing tiles and fittings – Product definitions and specifications*. There are also the following separate standards:

- BS EN 538 *Clay roofing tiles for discontinuous laying – Flexural strength test*.
- BS EN 539–1 *Clay roofing tiles for discontinuous laying – Determination of physical characteristics. Impermeability Test*.
- BS EN 539–2 *Clay roofing tiles for discontinuous laying – Determination of physical characteristics. Test for frost resistance*.
- BS EN 1024 *Clay roofing tiles for discontinuous laying – Determination of geometric characteristics*.

The machine-making process begins with the crushing of the clay to render the lumps of clay from the quarry into smaller-sized particles. This is usually performed by 'squeezing' the clay lumps through a set of high-speed metal rollers. The crushed clay may then have water added and worked further to ensure thorough mixing and consistency is achieved. This is then put into storage heaps and allowed to stand. During this time, the pressure in the heap allows its water content to be distributed evenly.

To achieve the desired shape, the processed clay is fed into an extruder or press where the air is removed by the use of a vacuum chamber which removes air pockets. The de-aired plastic material is then forced through a shaped mouthpiece at high pressure, resulting in a formed ribbon of clay. Further forming takes place by cutter or die, producing the final product which is then stacked using refractory setters or cassettes ready for drying and firing.

A few days are taken up by drying the newly formed clay products. The water in the clay body is slowly removed by gradually increasing the temperature in the drying chamber or tunnel. A lot of care has to be taken here in order to ensure that the drying is at the correct rate which ensures the clay is correctly conditioned and ready for firing.

The dry tiles are then moved into the kiln for firing, at a top temperature of up to 1,130°C, which can take several days to complete. Slowly increasing and then decreasing the kiln temperature and pressure allows the clay material to change to a hardened durable state suitable for the tiles to meet the requirements of BS EN 1304.

3.5 Large format tiles

Large format tiles can be either concrete or clay and are based on traditional single-lap designs, but larger and generally with an interlocking profile. These tiles were developed for economy and ease of fixing and laying, which saves on installation time, and they are available in a variety of shapes, colours, finishes and textures.

Many of the shapes available are based on traditional pantile and double Roman designs, although the recent aesthetic preference for flat 'slate-like' roofs, has produced flat interlocking tiles that have thin leading edges and are laid broken bonded to replicate natural slate.

They can be used at a range of different roof pitches from as low as 12.5° up to 90°. This is dependent upon the tile profile and performance characteristics determined by the manufacturer and based on testing for driving rain resistance and long-term experience in use.

3.6 Pantiles

Pantiles are single-lapped tiles that are a familiar sight in continental Europe and are already popular in certain counties of the UK, but their popularity is growing steadily outside these traditional areas. Their introduction into the UK is a legacy of several centuries of trading during which pantiles were brought back from the Netherlands and Belgium as ballast in the trading ships.

They are renowned for their distinctive S-shape profile(s) that creates a series of rolls and troughs to produce an elegant roofscape. Whilst the traditional clay pantile is a single pantile with a single 'S' profile where the right-hand side edge is turned down and the left-hand side edge is turned up, a related modern derivative is the double 'S' profile.

Single and double pantiles can be either non-interlocking (just overlapping) as per the traditional single clay pantiles or interlocking. They are made of either clay or concrete. In general, without an effective anti-capillary design, they should be laid at not less than 30° roof pitch.

With a single clay pantile, the top right-hand corner and the bottom left-hand corner are cut to a chamfer equal to the head-lap and side-lap (which prevents any gapping problems when laid). They include a nib and nail hole fixing moulded into the top edge on the back of the tile, at the bottom of the trough. The cut at the top right and bottom left is usually a straight line, but with some designs, it can be a zigzag or a curve. Provided all the tiles have the same corner shape, they will fit together. The cut starts along the top or bottom edge equal to the side-lap distance from the edge. The cut finishes down the side of the tile equal to the head-lap distance. The angle of the cut will be dependent upon those two dimensions.

3.7 Peg tiles and plain tiles

One of the oldest types of roof tile in the UK, the popularity of peg tiles made of clay grew due to the need for fire-resistant roof coverings, plus the simple rectangular shape (originally derived from wooden shingles) resulted in every town and village having its own artisan tile maker. The name 'peg tile', originally came from the wooden pegs, which were driven through two holes in the top edge of the tiles which allowed them to hang onto roof laths or battens.

Excavations show that some use of peg tiles in the UK dates back as far as Roman times but they were used extensively from the 12th and 13th Centuries. The colour, texture and shape of these popular regional tiles varied across the country depending on the mix of the local clay and aggregates, the temperature and length of time in the kiln as well as the skill of the tile maker. Peg holes were roughly rectangular, though circular holes became common in later years. By the mid-19th Century, clay peg tiles had evolved into 'plain' clay tiles which included nibs (or projections) on the underside of the top edge for hooking onto the battens. It is the plain tile that is nowadays more common, however, peg tiles retain an important but small sector of the roofing industry.

Plain tiles are still manufactured to the dimensions standardised by a Royal Charter of 1477 decreeing that all peg and plain tiles should be standardised to 10½ inch x 6¼ inch (267 mm x 165 mm). Peg tiles typically remained regionally sized with variations up to 1 inch (25.4 mm) in length and ½ inch (12.7 mm) in width.

Thanks to the plain tiles' small size, this design is ideally suited to traditional architecture and restoration work, but it is also frequently used on contemporary builds. Available in a variety of colours and textures, plain tiles can be handmade for a slightly irregular and rustic aesthetic, or machine-made for a cleaner uniform look. Subtle variations can also be seen between single-camber plain tiles (curved from top to bottom) and those that are double/cross-camber (curved from top to bottom and left to right). Both peg and plain tiles can also be formed incorporating an ornamental-shaped bottom edge such as arrow-head, bullnose, club and fishtail.

Peg and plain have no side interlocks, allowing water to enter through the gaps in between the tiles. To drain this water away safely, there must be another tile underneath each joint, hence the 'double-lap tiling' technique used when laying. They are also designed to be laid with a broken/stretch bond, with only the bottom third of each tile visible. This means that at any point on the roof there is at least a double layer of tiles, with three tile layers at the head-lap.

Plain tiles meeting the dimensional and geometric requirements of BS EN 1304 should be laid at rafter pitches not less than 35°. Some handmade tiles that do not comply with BS EN 1304 should be laid at pitches not less than 40°.

Handmade, crafted, machine made

Clay peg and clay plain tiles are manufactured in three basic ways – first by hand-making, second by hand-finishing (sometimes called hand-crafting) and third by machine-making-finishing.

'Handmade' tiles are made from clay taken from the lump, or prepared in the mass, from which a piece is cut, pressed into a simple mould and cut off, all by hand, with nibs formed and nail or peg holes formed using simple hand tools.

'Hand-finished' or 'crafted' tiles are made from clay prepared in the mass by machine and the tiles are then formed using automated machines. The finishing off is completed by hand to simulate an aged appearance similar to handmade tiles.

'Machine made' plain tiles are prepared without handwork by a variety of methods. Peg tiles will normally be, and in the past have been, produced by either the handmade or (latterly) the hand-finished process.

3.8 Reclaimed roof tiles

Using reclaimed materials for construction and interior design is increasingly popular due to the desire to create a rustic look. There is also a desire to upcycle items and use authentic materials from earlier time periods, to be environmentally friendly and sometimes cost effective.

Are new tiles always best?

With such a significant financial investment and crucial structural element as a roof, new tiles are determined by manufacturers and many builders alike to be always the best choice.

There are risks to reusing old tiles, as there are no guarantees on their performance, remaining lifespan or quality. There can be many hidden factors making a reclaimed roof tile roof unpredictable and potentially risky. Usually, there was a good reason that tiles had been removed from a roof.

Reroofing a house is a significant cost, so it is essential it is done correctly the first time. With new tiles, the occupant has peace of mind that they are guaranteed to last many years without repairs or replacement.

There are however times when reclaimed tiles are preferred to be reused. Historical and heritage roofs are a prime example, along with restoration work or building extensions where the property owner is looking to reuse existing tiles.

Quality reclaimed tiles can often be quite expensive as they are scarce and therefore a premium product. Some individual tiles may be in good condition, but a batch of old tiles would need to be carefully sorted through and checked to ensure that they are suitable for installing on a roof.

Thoroughly checking tiles is a very time-intensive process and imperfections may not be obvious, such as hairline cracks. Specialist 'heritage' roofing contractors are experienced in this aspect of their work, and there are some reputable reclaim yards who stock good reclaimed tiles which have already been checked and deemed fit for reuse. In both cases, the quality checking does add to the cost.

If there is an intention to use some reclaimed tiles on a project, these should not be intermixed with new tiles on a roof. Reused tiles will be easier to remove and replace should they start to fail if they are installed on a lower or separate elevation.

Creating the rustic heritage look

When it comes to achieving a heritage look with roof tiles on both historic and modern projects, manufacturers normally advise that it is better to weather new tiles, instead of old reclaimed originals.

Roof tiles in a variety of colours and styles are available, some of which have been designed with a weathered, rustic appearance, featuring irregular colour patterns and sanded textures for that 'faux heritage' aesthetic.

3.9 Fittings

Manufacturers offer a wide range of roof-tile fittings to complement their roof tiles. These can be concrete or clay, produced by hand or machine. Including traditional ridges, hips and valleys, which can be interlocking or separate, plus plain cloak verges and external vertical angle tiles, these are designed to aesthetically complement the roof tiles and provide practical weatherproofing to the roof. Individual valley tiles and hogback ridge tiles were made for use with peg tiles from the earliest times. In some areas, stone ridges were formerly used instead of clay.

Prior to the late 18th Century, hips were covered with lead sheet for security from wind. Individual round-pattern (or bonnet) hip tiles were introduced at this time in South East England (the Weald) and North Midlands (Broseley), being fixed to the hip rafter with iron nails as hooking with pegs would not have been secure.

Wider or narrower tiles used on row ends, could not normally be made, and gables were usually formed using single tiles, cut on site by the tiler. The gable (tile-and-a-half) tiles were not introduced until the beginning of the 20th Century. Angle tiles and half-round ridge tiles were then a later introduction.

Recent developments

Machine-made nibbed tiles dominated the plain tile market towards the end of the 20th Century due to mass production, and peg tiles became less popular being more expensive to lay.

Modern batten sizes now make it possible to nail tiles although there may be an issue with regards to nailing peg tiles into battens in that there are no nibs to take the weight of the tiles. This presents a theoretical risk of the battens splitting because of the forces placed on them, particularly at a steep pitch.

Present-day tilers have more experience with nibbed tiles than with peg tiles and a degree of specialist 'heritage' skills are needed when laying handmade peg or plain tiles. Tile manufacturers will always offer advice for laying their products.

The growth in demand for 'slate-like' roofs has encouraged machine-made tile manufacturers to produce a wide range of interlocking single-lap tiles that replicate the thinness of natural slate. With 'mock' vertical lines to represent smaller width slates together with a textured riven surface, they provide a realistic and more cost-effective solution to the natural product. This concept has also been used to replicate the smaller format of machine-made double-lap plain tiles, where larger interlocking flat tiles with thin edges provide an alternative to a double-lap plain tile, provide a realistic look and more cost-effective product.

CHAPTER 4

Accessories

4.1 Underlay

Roofing underlays

Roofing underlay, (also called membrane or felt), is installed under the exterior components of a roof, such as tiles or slates, and is designed to act as a secondary barrier against moisture in the roof space, as well as functioning in other ways to provide a weathertight roofing structure.

What is roofing underlay?

Low Resistance (LR) underlays are fabrics manufactured in three layers. Typically, the three layers are as follows:

- The top layer is UV-stabilised polypropylene to protect against the sun during installation and is usually the strongest of the layers.
- The middle layer is a microporous film which is water resistant and vapour permeable.
- The bottom layer is also made from polypropylene but is thinner than the top layer due to not needing the UV protection.

These layers contribute to the strength and weight of the product. Roofing underlays can be provided in a wide range of weights, typically, the heavier the weight of the underlay, the stronger it is, although this can depend on the actual composition. They are also easier and quicker to lay with the right amount of drape. This drape allows any rain ingress or condensate to drain safely down to the eaves avoiding the batten nailing positions on the rafters. Excessive drape should be avoided to prevent the transfer of wind loads to the external covering. Detail of wind load and wind uplift can be found in Chapter 2.3 *Wind uplift calculation*.

LR underlays are vapour permeable and airtight, having a vapour resistance below 0.25 MNs/g (Mega-Newtons Seconds per gram). There are also high resistance (HR) underlays, classified in BS 5250 as not being vapour permeable with a vapour resistance value above 0.25 MNs/g.

Air-permeable LR underlays, which allow for the transfer of both water vapour and air, require more onerous tile fixings when used.

The function of underlays

In addition to the secondary barrier function described earlier, one of the functions of a roof-tile underlay is to reduce the wind loading on the underside of the tiles. A roof will

DOI: 10.1201/9781003196990-4

experience both positive pressures and negative (suction) pressures. The suction pressures are usually referred to as wind uplift across different parts of the roof depending upon the roof geometry and pitch. The selection of the underlay should suit the expected wind loads acting on the underlay which serves to create a ballooning effect which can transfer the load onto the underside of the slates or tiles; therefore in higher buildings and locations that experience higher wind conditions, a stronger fabric is usually required. The wind resistance performance of all rolls of underlay should be clearly identified on the label in accordance with BS 5534.

An underlay should also, allow drainage of any moisture that might be deposited onto the surface of the underlay in the batten space and provide temporary weather protection before the installation of the primary roof covering.

Correct installation of any underlay should ensure that all horizontal overlaps in the underlay should be restrained to prevent the lap opening in windy conditions. This may be achieved either by coinciding the overlap with a tile batten or using an adhesive tape or integrated lap glue strips to secure the overlap

Any openings required through the underlay should be wide enough to receive spigots/pipes/vents with the cut flaps turned up in a manner so as not to allow any water penetration into the roof void, proprietary deflection trays should be used in conjunction with manufacturers' recommendations.

Those responsible for roof design should satisfy themselves that the roof construction as a whole is robust and any query relating to suitability can be confirmed by the underlay manufacturer directly. Designers should ensure compatibility with the proposed roof build up based on pitch, lap, ventilation requirements and suitability for the application and location.

General requirement

To be 'fit for purpose', the specification for a suitable pitched roof underlay, located beneath tiles and slates and draped over rafters, must satisfy the following performance requirements.

Roofing underlay should:

- provide a barrier to minimise the wind uplift load acting on the tiles
- provide a secondary barrier to the ingress of wind-driven rain snow and dust
- transport into the roof drainage system any moisture that may be deposited onto the surface of the underlay in the batten/counter batten space
- in the case of a type LR underlay, provide adequate water vapour transmission from the roof void into the vented batten space
- provide temporary weather protection (including rain impact resistance) before the installation of the primary roof covering (see the *UV resistance* section, next in this chapter)

Non-bituminous roofing underlay should be of suitable strength, water resistance and durability for the proposed application when tested to BS EN 13859–1, and this should be confirmed by the relevant UKAS Accredited Third Party Certification and by the underlay manufacturer.

When used without ventilation, the underlay should also have a water vapour transmission rate high enough to prevent the formation of condensation beneath the underlay, and will usually be subject to providing a sealed ceiling construction including a vapour control layer in accordance with BS 5250.

Ultraviolet (UV) resistance

The pitched roof underlays are designed as secondary water-resistant membranes and should not be designed as the primary waterproofing layer.

An underlay which is left exposed for any period of time is subjected to ultraviolet (UV) light (sunshine), which may lead to premature failure during its service life. Therefore, the exposure period should be kept to a minimum.

If severe weather conditions prevail, the underlay should not be left exposed for more than a few days. In the event that it has to be left without a roof covering for some time, then a tarpaulin or other similar protective sheeting should be used to protect it until the roof covering is completed. In all cases, it is important to ensure that the primary roof covering is fixed in place as soon as possible.

The material/product which is exposed as the drip into the gutter must be adequate to remain functional during its design life, however, there are no standards for assessing this performance. Specialist products for this purpose such as a proprietary Underlay Eaves Carrier Strip or 330 mm wide strip of BS 8747 Type 5U reinforced bituminous underlay are recommended.

Workmanship requirements

Handling/laying

The material should be able to be laid under normal roof climatic conditions without trauma and the size of the roll used should optimise both the transportation to the roof and the speed of laying.

Fixings

Clout nails, no less than 3.00 mm shank diameter are preferred to staples for fixing, and these should be kept to a minimum to reduce the risk of water penetration. Some flexible underlays both HR and LR can extend around the fixings when subjected to temperature changes and wind.

The use of clout nails allows a better degree of self-sealing, as previously experienced with 1F bituminous felt.

Where practically possible, fixings should be under laps to minimise penetrations. Proprietary products are available for sealing fixing penetrations in non-bituminous underlays.

Application advice

Drape

Underlays may be laid draped over rafters or counter battens at recommended spacing suited to their tensile strength. The drape should be sufficient to allow drainage of water away from the batten fixing holes directly above the rafters or supports and beneath the battens (confirm using a controlled water test if necessary). At its slackest, the drape should be no more than 15 mm, in order to conform to BS 5534 and BS 8000–6 and minimise the risk of ballooning and contact with the underside of the roof covering.

Where underlay is found to be draped in excess of 15 mm, the individual manufacturer should be contacted with regards to its performance and any issues relating to warranties or guarantees.

Secured laps

All underlay laps must be secured in some way by a batten or proprietary means whether the roof underlay is fully supported or unsupported. Where the laps are to be secured by battens, there are two methods of complying with the requirements of BS 5534 and BS 8000–6.

The first is to increase the horizontal laps where necessary, to ensure they coincide with a regular batten course. Using this method, more underlay may be used, but no additional battens are required, and trip hazards are avoided.

The second is to maintain consistent underlay laps and, if necessary, to install additional battens over them. It should be noted that if additional battens are used, they should be of the same size and quality as those used for the roof covering.

Laps can also be secured using tapes provided by the manufacturer, or with glue strips integrated into the edge of the underlay.

Counter battens below and above the underlay

Where a roof has counter battens fixed below the underlay, it is important to ensure that the drape of the underlay does not reduce the minimum air gaps required by BS 5250. Where a roof has counter battens fixed above the underlay, it is not necessary to drape the underlay and battens, and laps are not required although taped/glued laps may still apply if specified.

Temporary covering

Underlays may be used as a temporary covering over continuous sarking boards, remaining in position after the fixing of slates, which are nailed through the underlay to softwood boarding, as in Scottish practice. Where this is the case, temporary battens are often fixed in place to secure the underlay, until the roof is tiled.

Minimum laps

The minimum unsealed head-laps for underlays are shown in Table 4.1 as follows:

Table 4.1 Minimum underlay laps

Minimum underlay laps		
Rafter pitch degrees	**Minimum head-lap**	
	Not fully supported (mm)	**Fully supported (mm)**
12.5 to 14	225	150
>15	150	100

With side-laps, the minimum lap for underlay is 100 mm for all roof pitches. At all other junctions on the roof, including jointing to the main roof, the underlay from one side should overlap the other by no less than 150 mm beyond the junction line (for example, hip or common rafter).

Proprietary adhesives and tapes are available for sealing lap joints of underlays, and if used should be installed in accordance with the manufacturer's recommendations. The minimum sealed laps are usually not less than 100 mm.

Where an underlay overlap does not coincide with the batten, consideration should be given to either including an extra batten at the overlap or increasing the underlay lap to coincide with the next batten.

All penetrations, pipes, vents, etc. through the underlay should be suitably detailed to prevent water ingress. Except at ridges, where small openings above the top batten are permitted, devices that open the laps of the underlay should not be used.

Bats

Recent studies by Reading University have found that modern flexible lightweight underlays are detrimental to bats because their claws get caught in the material. Whilst the main issue is obviously the welfare of the bats, the potential damage to the underlay should also be considered. The research demonstrates that traditional under-tile reinforced bitumen felt is safe for bats.

It is illegal to damage or destroy a bat roosting place (even if bats are not occupying the roost at the time), therefore it is strongly recommended that bitumen felt is used for all roofs wherever there is evidence that bats may be roosting, or where they are likely to roost (for example, textured stone/slate roofs with obvious gaps in the coverings). It is important that contractors are aware of how to proceed without impacting the bats and it is therefore recommended that they seek specialist advice when working on roofs where they are or are likely to be present.

Bat-related enquiries including advice regarding breathable roofing membranes and their impact on bats can be found on the Bat Conservation Trust website, www.bats.org.uk.

Bat access tiles and systems are available from many tile manufacturers and may be a consideration to include in the roof covering design.

Building regulations (alternative approaches)

To allow for innovation and proprietary systems, compliance with certification from a UKAS-accredited body will, in some cases, meet the requirements of The Building Regulations, or third-party warranty providers such as NHBC or similar. Users and designers are advised to check the requirements of any such certificates carefully to consider the whole building, and to check with individual manufacturers before proceeding.

Summary

It is important to fully consider all the performance factors and workmanship factors involved in the use and function of any pitched roof underlay. This will enable the user to quantify and qualify the suitability of the underlay. Such a comprehensive performance and workmanship assessment is particularly necessary when traditional products or roof details are changed, which may have an intrinsic effect on their fitness for purpose.

More information on underlay is given in Chapter 6.1 *Laying the underlay*.

4.2 Ventilation products

As highlighted in Chapter 2.8 *Condensation in roof spaces* and Chapter 2.9 *Ventilation in cold and warm roofs,* changing lifestyles along with the increased use of insulation and draft proofing in buildings has led to an increase in condensation within roof spaces. BS 5250 requires all roofs to be ventilated in order to control condensation, and roof-tile manufacturers have responded to this by producing a matching range of accessories which facilitate this ability to ventilate.

The degree to which the roof needs to be ventilated depends on several factors, and this is covered in previous sections. This section looks at some of the commonly available ventilation products.

Eaves vents

This method of ventilation usually comprises a continuous ventilator strip attached to the top of the fascia board, allowing air to follow into the roof space underneath the tiles and the under-tiling felt. This over-fascia ventilation provides a discrete method of ventilating the roof space as it is usually out of sight behind the guttering.

Over-fascia vents should always be used in conjunction with eaves support trays, and if the insulation is at ceiling level, spacer trays or rafter trays should be installed in order to prevent the insulation from blocking the passage of air into the roof space. Quite often these three items are available in kit form with 10 mm and 25 mm options to suit cold and warm roofs respectively.

The depth of fascia vents should be considered when setting the height of the fascia board, if not, a tilt at the eaves could lead to a flat spot or risk of ponding. Due to the fact that they form a continuous air gap along the eaves, over-fascia vents are an effective method of ventilating a warm roof.

Felt support tray

Felt support trays are usually fitted above the over-fascia vent and are designed to adequately support the roofing membrane to prevent ponding at the eaves, as well as prevent damage to the roofing membrane at the eaves.

Eaves comb filler

Eaves comb filler is installed above the felt support tray when profiled tiles are being used. The eaves comb filler effectively prevents the entry of birds, rodents and large insects from entering the batten cavity.

Continuous rafter tray

Continuous rafter tray or roll-out rafter tray is a simple but effective way of preventing flexible quilt loft insulation from being pushed up tight to the underside of the roofing

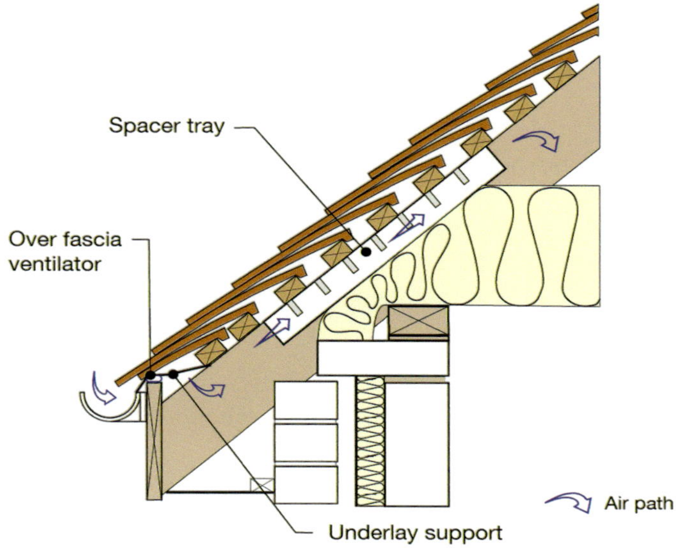

Figure 4.1 Eaves vent

underlay. This creates and maintains a clear ventilation path into the roof void from the eave's ventilation.

Dry ventilated ridge & hip systems

BS 5534 requires all ridge tiles to be mechanically fixed, and manufacturers have responded to this by designing systems which meet this requirement whilst also incorporating the 5000 mm²/m of high-level ventilation as required by BS 5250. Some of these systems use a ventilated strip on either side of the ridge, whilst others use a ridge roll which is laid underneath the ridge tiles to seal the roof from the elements whilst incorporating ventilation.

In most cases, the ridge tiles are secured using metal screws and clamping plates, which fix into the ridge batten. These systems are designed to be used in conjunction with most standard ridge tile profiles, although this can vary from one manufacturer to another so compatibility should be checked prior to installation.

Currently, there are no British or European Standards for roof ventilation products, however, due to the increased use of these systems over recent years, a new standard, BS 8612 has been introduced, which sets out key minimum performance and durability criteria of dry fix fittings and accessories along with test methods. These essential requirements include:

- Durability and material specifications.

Roof ventilation products must be manufactured to remain durable under normal exposure conditions and without loss of performance for the stated design life of the product. Colourfastness would normally be excluded from the manufacturer's durability and performance guarantees.

- Mechanical resistance (strength and security)

Roof ventilation products are not normally designed to withstand pedestrian loads. Product strength and that of its fixings must be adequate to withstand anticipated wind and snow loads for its design life.

- Rain performance

Roof ventilation products must adequately resist the ingress of driving rain and snow both through the product and by its integration with surrounding tiles and slates. The wind-driven rain and snow performance between the product and surrounding tiles and slates must be at least comparable with that of the tile or slate array. The product manufacturer should declare details of the test method, test parameters and the product's performance.

- Ventilation characteristics

The airflow resistance of the product, measured at defined airflow rates, along with the geometric free airspace, measured at the product aperture's narrowest point, should be declared by the manufacturer.

- Installation instruction

Ventilation products must be installed appropriately to provide the required level of ventilation as described in the manufacturer's technical information. It should be noted that when used with flat tiles, some systems provide little ventilation due to the restricted airflow underneath the ridge tiles, and the manufacturer's specifications should always be checked to ensure that ventilation is sufficient.

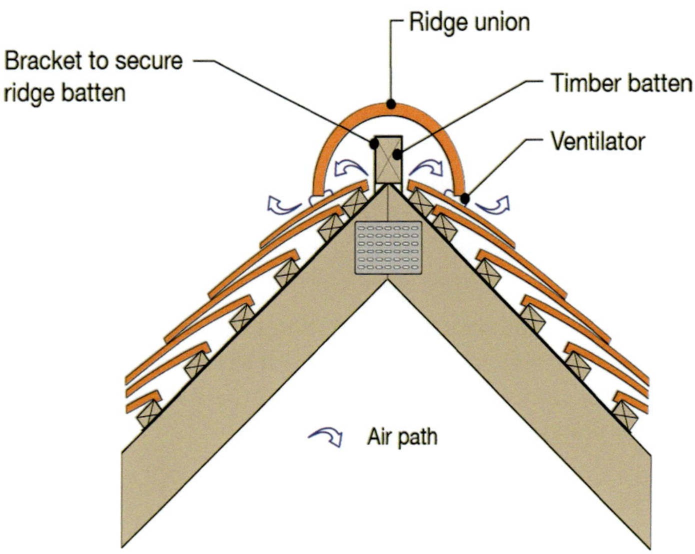

Figure 4.2 Dry ventilated ridge system

Examples of ventilated dry ridge systems

Figure 4.3 shows an example of a proprietary ventilated dry ridge system using ventilated strips on each side of the ridge to maintain a ventilation gap at a high level. Plastic joints provide a waterproof seal between the ridge tiles, which are secured using screws or nails and small metal plates.

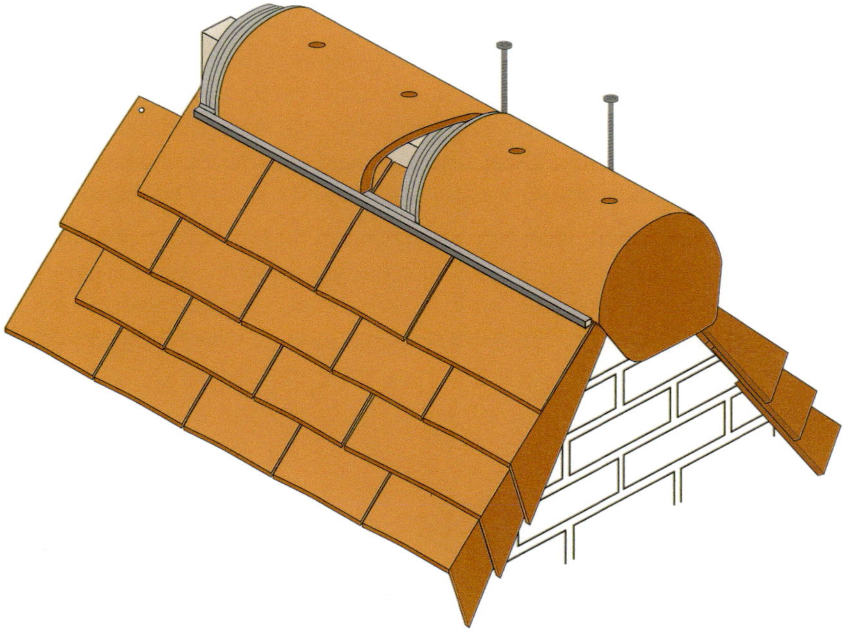

Figure 4.3 Ventilated dry ridge system using ventilated strips

Figure 4.4 shows an example of a ventilated dry ridge system using a universal ridge roll. The ridge roll provides high-level ventilation as well as a seal at the ridge. In this example, holed ridge tiles are secured using nails or screws with sealing washers.

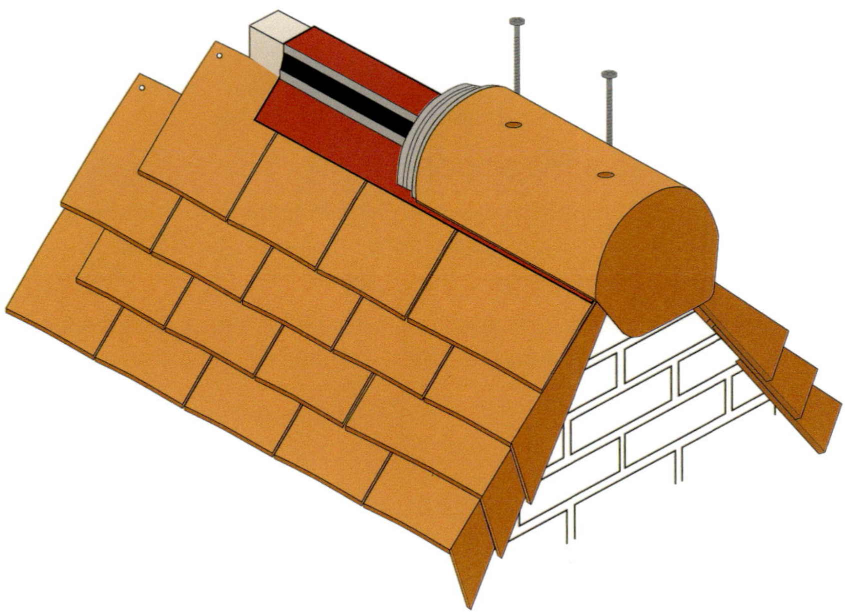

Figure 4.4 Ventilated dry ridge system using ridge roll

Tile vents

These can be used for both roof space ventilation and as a mechanical extract terminal using an additional adaptor kit. Roof-tile manufacturers design these vents to match specifically with each tile in their range. The ventilation capacity of these vents varies from one product to another and should always be checked prior to installation. Tile vents may be limited by roof pitch and manufacturers recommendations should be followed.

When ventilating the roof space, tile vents can be used for low-level or high-level ventilation when installed at the appropriate frequency to meet the ventilation requirement. For example, tile vents which provide 7500 mm^2/m of ventilation would need to be installed every 0.75 m to give 10000 mm^2/m ventilation at low level or every 1.5 m to give 5000 mm^2/m of high-level ventilation.

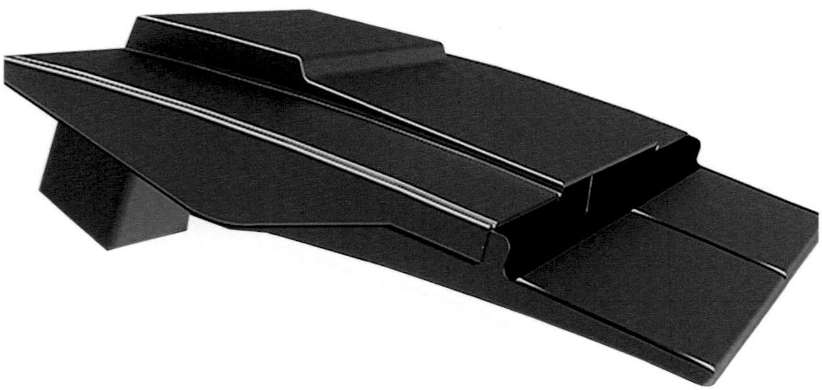

Figure 4.5 Example of a plain tile vent (*Hambleside Danelaw*)

Ridge tile vents & gas vent ridges

Ridge tile vents are produced to match the ridge tiles in each manufacturer's range, however they are less commonly used due to modern dry-fix systems. If available, they can be used for both roof-space ventilation or mechanical extraction.

Gas vent ridges are available for terminating gas flues which are usually connected to gas fires, these are becoming less popular. These should only be connected by qualified installers.

Abutment vents

These are designed to provide ventilation where the roof slope butts up to a wall, for example, where the roof meets the bottom of a dormer or where a lean-to roof meets a wall at the top. These situations require high-level ventilation of 5000 mm^2/m as per BS 5250.

Abutment vents usually comprise a continuous vent strip which sits beneath the flashing, providing a discrete means of allowing the air to flow out of the roof space.

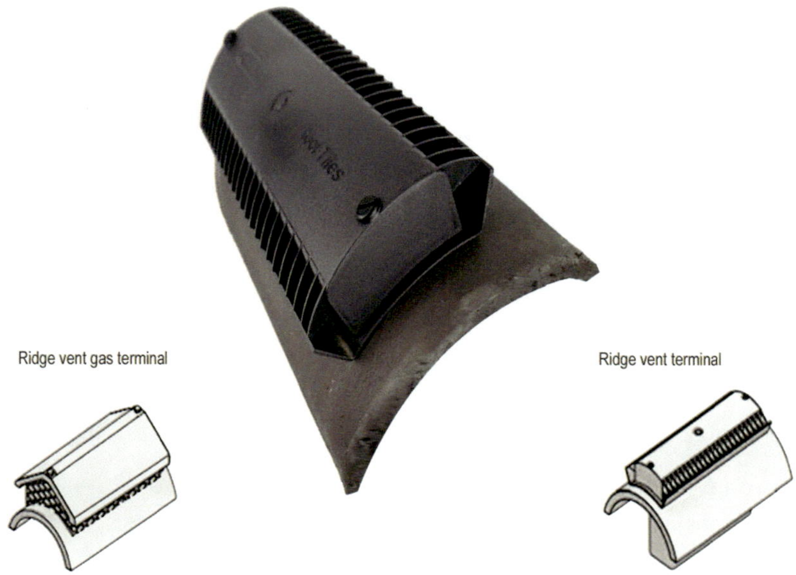

Figure 4.6 Examples of a ridge tile vent (*Marley*)

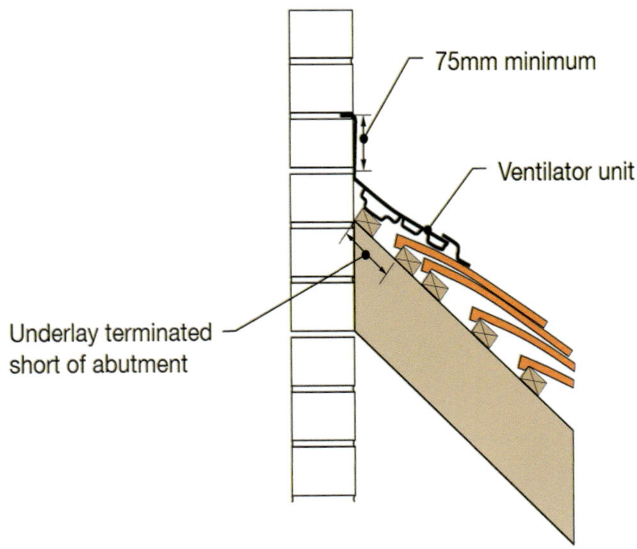

Figure 4.7 Abutment vent

4.3 Ridges, hips, finials, verges

Fixing of ridges and hips

The traditional method of fixing ridge tiles (and ridge tiles down the hips) was to bed them using mortar only. Hips made use of a hip iron to prevent slide.

In recent years, largely driven by severe weather causing increasing insurance claims for ridge repairs, changes to BS 5534 were made to raise standards with the installation and performance of roof tiling. These changes identified the need to mechanically fix ridge and hip tiles as the use of mortar only was found to be unreliable. This led to the increased use of dry-fix systems.

Although BS 5534 recommends mechanical fixing on new roofs and replacement roofs (including repairs), it acknowledges that in the case of heritage roofs, its recommendations may not be appropriate. Traditional fixing methods for heritage roofing materials often conflict with BS 5534, and consultation with local planning authorities and conservation experts (including specialist roofers) is advised so that a suitable approach can be agreed.

Further details can be found in Chapter 5.4 *Roofing mortar*.

Common examples

Ridges, hips, cloaked verge and finials fitting tiles are specialist tiles that can serve functional and decorative requirements. There are many different types, and this chapter identifies some of the more common products.

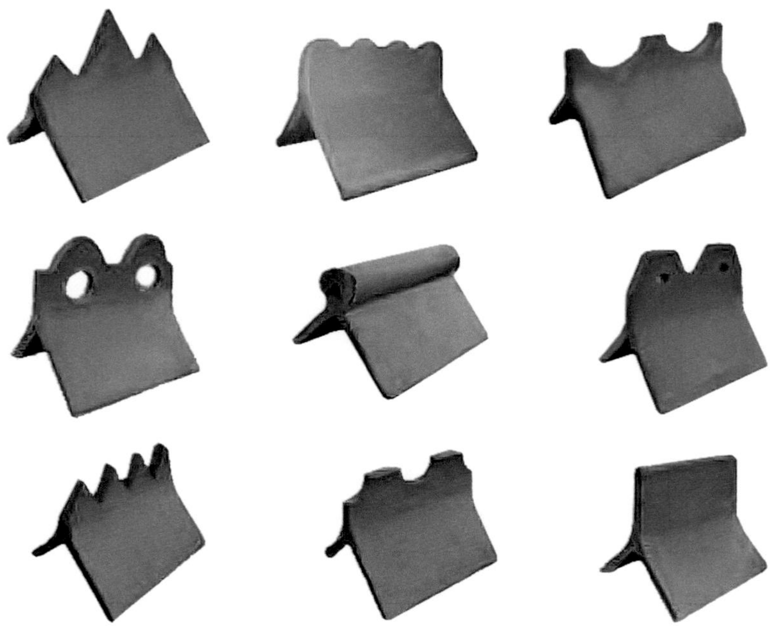

Figure 4.8 Some examples of decorative ornamental ridges

Gable stop end

Figure 4.9 shows an example of a gable stop end ridge tile. These are commonly used at the gable in conjunction with bedded verges and avoid the need for a deep bed of mortar underneath the end ridge tile. All ridge tiles must be mechanically fixed to the roof structure in accordance with the manufacturer's instructions.

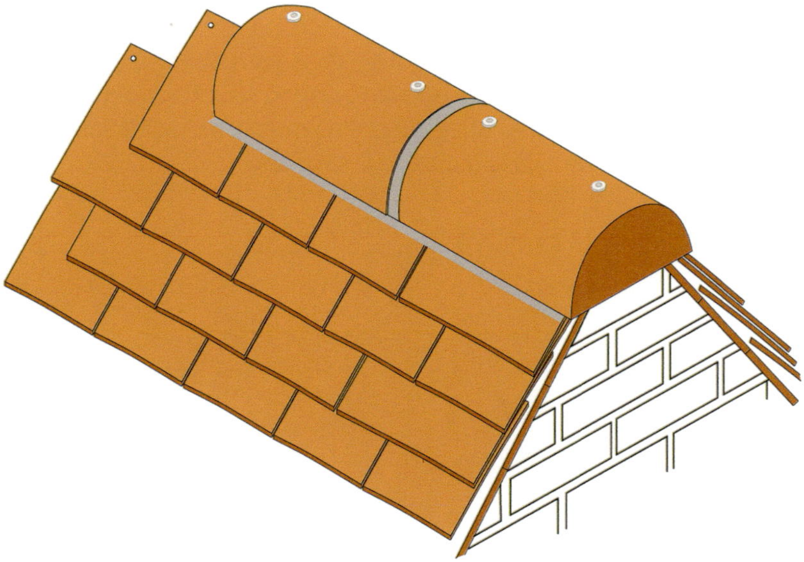

Figure 4.9 Gable stop end

Block end

Drawing 4.10 shows an example of a gable block end ridge tile. These are commonly used at the gable in conjunction with cloaked verge tiles or dry verge systems, although they can also be used with bedded verges to avoid the need for a deep bed of mortar underneath the end ridge tile. All ridge tiles must be mechanically fixed to the roof structure in accordance with the manufacturer's instructions.

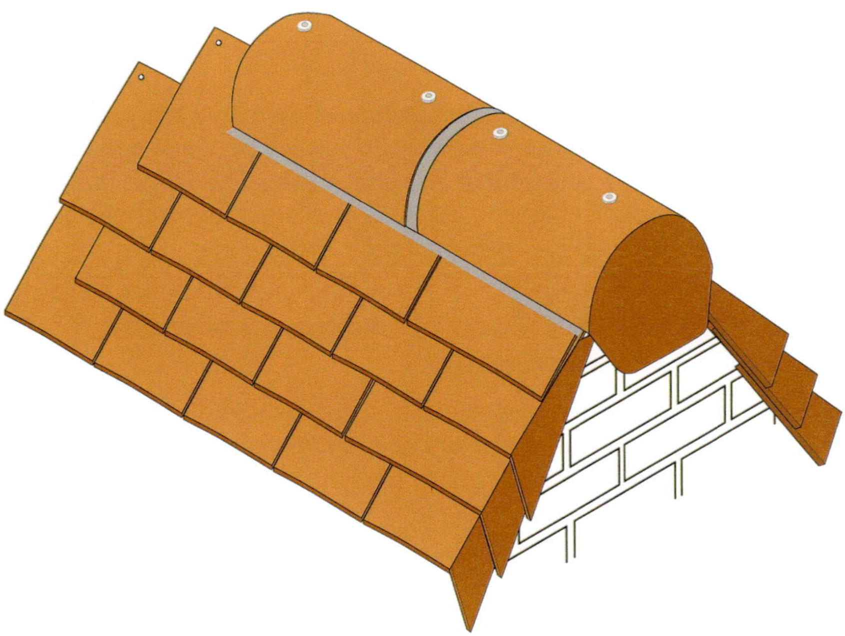

Figure 4.10 Gable block end

Hip starter

Figure 4.11 shows an example of a hip starter tile. These are commonly used at the eaves on a dry hip, although they can also be used with bedded hips to avoid the need for a deep bed of mortar underneath the end hip tile. All hip tiles must be mechanically fixed to the roof structure in accordance with the manufacturer's instructions.

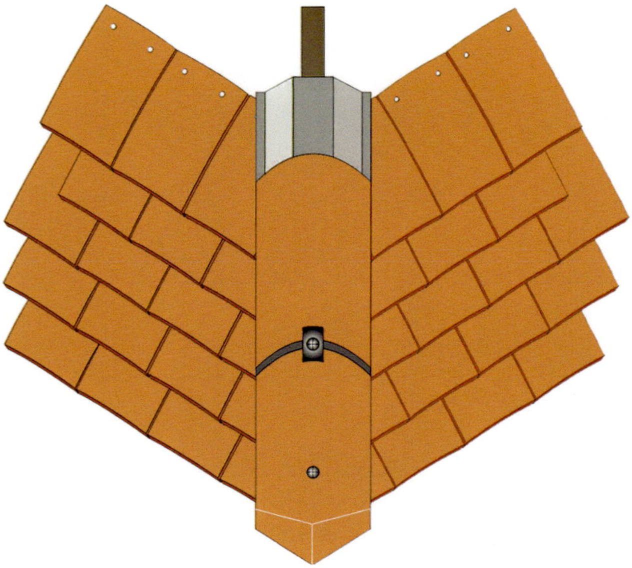

Figure 4.11 Hip starter

Hip ridge junction fittings

Figure 4.12 shows an example of a hip ridge junction. These are used predominately with plain tiles at the junction between ridges and hips on which bonnets, arris hips or a mitred hip have been used. A suitable lead saddle (not shown) should be fixed underneath the end ridge tile to weather the ridge/hip junction. All ridge tiles must be mechanically fixed to the roof structure in accordance with the manufacturer's instructions.

Mechanically fixed ridges – use of nails or screws or proprietary manufacturer fixings

Figure 4.13 shows an example of a mechanically secured ridge tile using nails or screws with sealing washers. The ridge tile must be drilled or specially manufactured with holes. Where the ridge tree is absent, or of insufficient height or width to accommodate the fixings, it will be necessary to fit an additional ridge timber.

When using this method of mechanically fixing in conjunction with mortar bedding, it is better to use screws rather than nails to avoid the risk of dislodging the mortar during fixing. Some manufacturers supply proprietary fixing kits to mechanically fix bedded ridge and hip tiles.

Figure 4.12 Hip ridge junction

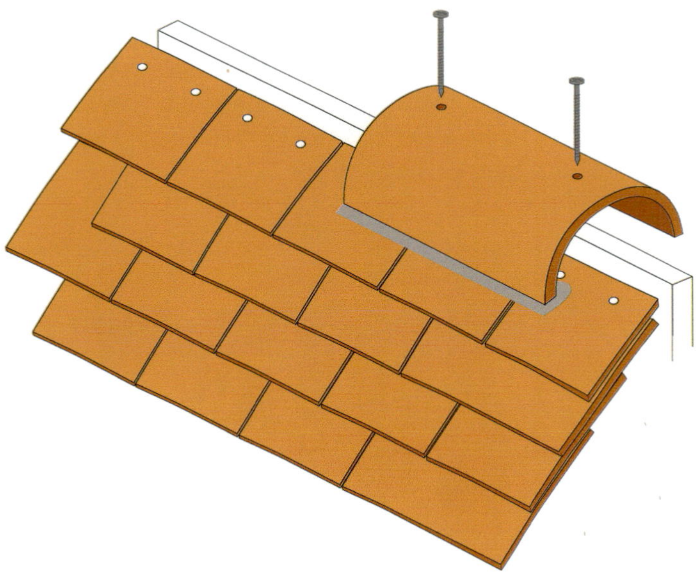

Figure 4.13 Mechanically fixed ridge

Decorative finials

Decorative finials provide a similar function to the specialist tiles shown above, but they are accentuated with decorative features to give the building a sense of grandeur and craft. Decorative finials are handmade by skilled craftsmen.

Figure 4.14 Examples of decorative finials (*Dreadnought Tiles*)

Mechanical fixing of finials

Due to the weight and height of finials extra care and attention should be given when fixing them to ensure stability, please follow the manufacturer's recommendations. Most finials have a specific fixing method, in this example there is a hole to enable it to be nailed or screwed to the ridge batten using a nail or screw with a sealing washer. Other fixing methods can include threaded bars or metal straps.

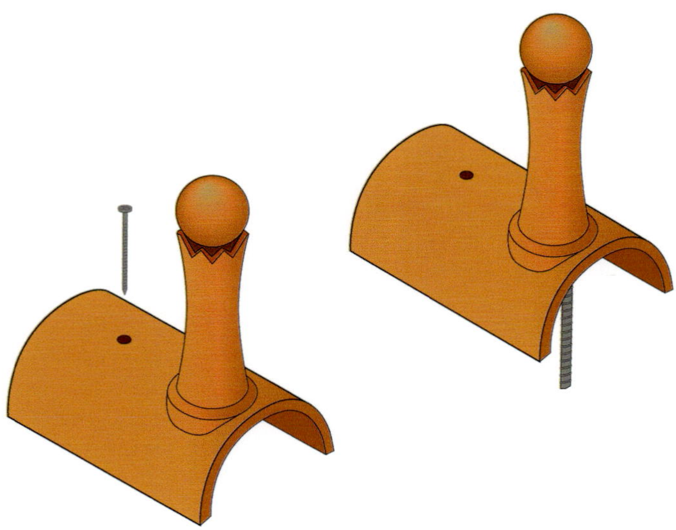

Figure 4.15 Mechanical fixing of finials

Cloak verge tiles

Cloak verge tiles are purpose-made and colour-matched tiles that provide a strong, weathertight and maintenance-free dry verge whilst matching the existing tiles on the roof. Cloak verge tiles meet the requirements of BS 5534 as well as providing a mechanical fixing at the verge of the roof.

Dry fix verge systems

Dry fix verge systems are products that cloak the roof covering and close any visible gaps between the tiles and either the gable end wall or bargeboards, providing a neat and aesthetically pleasing finish on the verge. They usually provide a second mechanical fixing for the tile as well as preventing any water or pest ingress into the roof space whilst diverting any water away from the roof covering.

Dry fix verge systems are required to meet BS 8612 standard and come in a variety of forms. Usually manufactured from plastic or aluminium, they can be individual interlocking units or continuous lengths.

4.4 Undercoating

Originally, the only recognised roof undercoating was sand lime mortar reinforced with animal hair applied to the head-laps of double-lapped slates or tiles. This system was commonly known as ‘torching’ and was used before the introduction of roofing underlay to increase wind-proofing and resistance to wind-driven rain and snow in exposed locations. Although BS 5534 states that ‘torching’ is undesirable for standard roofs, it is important to recognise that the practice still forms an essential part of some conservation or heritage specifications.

In recent times, a series of roof undercoatings have come to the market, particularly spray foams. Initially marketed as a cheaper and quicker alternative to a complete reroof with the benefit of added insulation, the application of a spray foam roof undercoating installed incorrectly by non-accredited tradespeople has created numerous issues. These undercoating practices are not recognised or endorsed by roof-tile manufacturers as a stabilisation method for roofs due to the reports of associated defects such as condensation and timber rot.

Warranty providers such as LABC Warranty or NHBC will not accept any spray foam insulation (even products which hold third-party accreditation such as BBA) for application to pitched roofs. Mortgage lenders are more wary of lending on roofs treated with these types of undercoat systems, and the use of spray foam may invalidate home insurance policies.

CHAPTER 5

Materials handling and preparation

5.1 Health and safety

Falls from height remain one of the biggest causes of accidents and fatalities in construction, therefore, it is important that the hierarchy of fall protection set out in the Work at Height Regulations 2005 is followed at all times.

Home builders constructing new homes should install internal fall protection to protect the workers installing roof coverings from injuring themselves whilst working above open roof trusses. Types of internal fall protection include proprietary decking systems, netting or airbags/bean bags as soft-landing systems.

Following an investigation, it was noted that some UK house builders allow internal fall-arrest systems to be removed once the roof is felted and battened, unfortunately though, this only addresses one aspect of the risk involved. Although operatives are instructed and trained to only walk where the tile battens are attached to the trusses, it is possible for the operative to lose their balance and place their foot midspan of the truss to prevent them from falling over.

This risk increases further when the roof is being loaded out with roof tiles because the operative is standing in the upright position along the truss line whilst walking up the roof carrying the roof coverings. In the event that a roof batten breaks under the weight of an operative, there is only the strength of the underfelt membrane to prevent them from falling through into the roof void, and as such, the potential risk of injury is still high.

It is therefore recommended that the internal fall protection is not removed by the home builder or main contractor until it is safe to do so. This would be ideally when the roofing works have been completed, but as a minimum, when the roof has been felted, battened and completely loaded out. If it is necessary to remove the fall protection before this work is completed, a risk assessment should be carried out to show that it is safe to do so.

The recommendations in the HSE guidance note HSG 33 – *Health and Safety in Roofwork* and in HSG 150 – *Health and Safety in Construction* should be followed.

This section provides guidance on safety method statements, a list of legislation that may apply, risk analysis and guidance on roof access. However, please refer to the HSE website for the most up-to-date information:

Safety method statement

A safety method statement should be prepared that includes:

DOI: 10.1201/9781003196990-5

- all working positions and access routes to and on the roof
- how falls are to be prevented
- how the danger from falling materials to those at work and to the public is to be controlled
- how risks to health will be controlled
- how other risks identified at the planning stages are to be controlled
- what equipment will be required
- what competence and training will be needed
- who will supervise the job on site
- how changes will be made to the work without prejudicing safe working
- who will monitor that the safe system of work is operating properly

Legislation

With particular reference to roofs, the following laws could apply. However, please refer to the HSE website for the most up-to-date information:

- ***Health and Safety at Work Act (HSW) 1974*** – Applies to all work employers, employees and the self-employed.
- ***Management Health and Safety at Work (MHSWR) 1999*** – Applies to all work employers, employees and the self-employed. Assess and reduce risks.
- ***Construction (Health, Safety and Welfare) 1996*** – Applies to all construction work employers, employees, the self-employed and all those who can contribute to the health and safety of a construction project.
- ***Construction Regulations 1989*** – Applies to all requiring head protection.
- ***Construction (Design and Management) Regulations 2015*** – Sets out what people involved in construction work need to do to protect themselves from harm and what actions are required to deliver building and construction projects in a way that prevents injury and ill health.
- ***Lifting Operations and Equipment (LOLER) 1998*** – Applies to all lifting equipment.
- ***Manual Handling Operations 1992*** –Applies to employers and the moving of objects by hand or bodily force.
- ***Provision and Use of Work Equipment (PUWER) 1998*** – Applies to all equipment providers including machinery which should be safe for work.
- ***The Workplace (Health, Safety and Welfare) 1992*** – Applies to employers regarding ventilation, heating, lighting, workstations, seating and welfare facilities.
- ***Health and Safety (First Aid) Regulations 1981*** – Provision of suitable first aid facilities and at least one trained first aider.
- ***Reporting of Injuries, Diseases and Dangerous Occurrences Regulations (RIDDOR)*** – Requires employers to notify certain occupational injuries, diseases and dangerous occurrences.
- ***Noise at Work Regulations 1989*** – Requires employers to take action to protect employees from hearing damage.
- ***Electricity at Work Regulations 1989*** – Requires people in control of electrical systems to ensure they are safe to use and maintained in a safe condition.
- ***Control of Substances Hazardous*** *to Health Regulations (COSHH) 2002* – Requires employers to assess the risks from hazardous substances and take appropriate precautions.

- *Personal Protective Equipment at Work 1992* – Applies to employers for the provision, use and storage of appropriate protective clothing and equipment.

Risk analysis

The *Management of Health and Safety at Work Regulations* require employers to carry out risk assessments, make arrangements to implement necessary measures, appoint competent people and arrange for appropriate information and training. This requires:

- looking for hazards
- identifying who and how they can be harmed
- evaluate and action to eliminate or reduce risk
- recording or communicating findings
- review findings

When looking for hazards, consideration should be given to:

- trees
- overhead phone and electric cable
- large eaves overhang reducing the stability of access equipment
- vulnerable building features such as plastic gutters and castellated ridge tiles
- aerial, satellite dish and phone cables laid across the roof
- uneven roof rafters
- doorways or other pedestrian routes below work areas
- weak roof structures below work areas such as conservatories
- uneven or soft ground below work areas
- length of roof pitch to suit available roof ladders
- clearance of hidden roof structures such as flashing or aprons from chimneys or valleys
- features of the building enabling tie-in of access equipment such as opening windows
- routes of electric extension cables avoiding the risk of tripping
- overlooking private areas/screening

Personal access and working at heights

A suitable working platform at or near the eaves level is an essential safety item when working above 2 m height. Working platforms feature:

- minimum width to cover the working area whilst on the roof
- sufficient strength to support loads including personnel, tools and materials
- sufficient integrity to stop tools and debris falling below
- main guardrail at least 910 mm above fall edge
- toe-board at least 150 mm high
- no unprotected gap between these exceeding 470 mm (i.e. use a mesh of intermediate rail)

Batten safety

Battens supplied and fixed in accordance with BS 5534 are designed to have adequate strength to support the dead weight, imposed load and wind load on a roof clad with tiles (or slates).

Pre-graded 25 mm × 50 mm battens that meet BS 5534 grading are at least 1.2m long to ensure they span a minimum of three trusses, and are fixed with the recommended nails, may be used as an alternative to roof ladders in line with current guidance in the Health and Safety Executive's publications HSG 33 *Health and Safety in Roof work* and INDG 401 *Working at Height*. At all times, only trained tilers should have access to any roofing assembly of underlay, battens or tiles during the tile-covering construction process.

The safe place of work for the tiler will usually be on, or supported by, the line of the rafters. Workers should not stand directly on unsupported battens or tiles so as to avoid unnecessary risks or damage. Any existing tiled roof should be treated as a fragile roof unless a competent person has inspected it and declared it non-fragile.

The use of ungraded battens can put the safety of the operative at risk, which can put the contractor at risk for breach of his 'duty of care'.

Safety

For all roofs of standard construction, it should be assumed that the roofing works will need to be in accordance with BS 5534. Therefore, the battens should conform fully to BS 5534, with regards to grading, marking, treatment and documentation.

Some organisations, such as the NHBC, expressly state that battens should arrive on site fully graded to BS 5534 (showing the factory grading). Failure to comply with this requirement may risk invalidating certain insurances or warranties. All supplies of fully graded battens should be marked as complying with BS 5534 and each delivery should be accompanied by documentation to demonstrate full traceability. It is also recommended that battens should be manufactured with an independently assessed product quality-assurance system, either UKAS accredited or other recognised third-party certification scheme.

It is generally accepted by the industry that whatever method of grading is used a small percentage of incorrect grading is likely, although there is no allowance for this in BS 5534. It is recommended that the permissible incorrect grading rate for battens used in the UK should be no more than 10% for borderline deviations and 0% for clear or significant failures.

Users and purchasers of battens should be aware that low failure or incorrect grading rates are an important indicator of quality and risk reduction. All suppliers of graded battens should therefore strive for as low a failure rate as possible and maintain a robust grading system.

Whilst factory-graded battens have been checked at source, no grading process can be 100% accurate and damage can occur in transit and/or during the works. Therefore, it is recommended that even fully graded battens should be checked before installation for obvious defects and also split ends or damage that may have occurred after manufacture.

Batten sizes

BS 5534 includes batten sizes to meet the concerns expressed regarding stiffness (batten bounce) and adequate fixing to the battens. The recommended batten sizes include the limits of the permitted tolerances as well as providing adequate strength in service.

The thickness is now an absolute minimum thickness when measured at a 20% moisture content. A small dimensional allowance for timber shrinkage and swelling should be made when battens are measured at a moisture content less or greater than the reference moisture content of 20%. The moisture content at the time of fixing should not exceed 22%.

Under no circumstances should batten sizes less than those recommended to be used, always with adequate assurance that the batch of battens fully complies with the revised grading rules.

In the UK, battens dimensions are 25 mm x 38 mm and 25 mm x 50 mm. The choice of the correct size to use will depend on several factors, including span, type of roof covering and whether the tiles are laid with a single or double-lap. For double-lap clay and concrete tiles, the correct timber batten size is 38 mm x 25 mm. Single-lap installations should use 38 mm x 25 mm battens for 450 mm rafter spans and 50 mm x 25 mm battens for 600 mm rafter spans.

Preservative treatment

It is recommended that a timber treatment certificate is obtained for each batch of battens. The moisture content of the timber pretreatment must not exceed 20%, otherwise the treatment can be ineffective. Failure to comply would invalidate or compromise any treatment warranties (typically a 60-year lifetime guarantee).

Cut batten ends, especially when in contact with mortar, should be treated with a suitable preservative approved for this purpose as per the requirements stated in BS 5534 Annexe E.

Identification, marking and documentation

Each delivery of battens should be accompanied by documentation stating at least the following:

- Name of supplier.
- Origin/timber species type.
- Graded in accordance with BS 5534.
- Basic size or sizes.
- Type of preservative and method of treatment if applicable.

Each batten must be clearly and indelibly marked in accordance with BS 5534 clause 4.11.5 to indicate the supplier, origin/timber species type, grade and size. In addition, it is recommended that the quality mark of the assessing body should be included. Whilst most factory-graded battens are coloured for ease of identification, it should be noted that this is not a requirement under BS 5534.

5.2 Estimating materials

In this section, we will describe a typical approach to estimating quantities of roofing materials. Often referred to as a 'take-off', it is a method to assist with assessing the size of a job, allowing you to order all the necessary components and to provide a quotation for your customer and job.

Estimating quantities for a specified roof tile should always be carried out with reference to the technical data issued by the specific tile manufacturer, such as the number of tiles per square metre coverage, ridge tile lengths and so on.

Tools required

- Protractor.
- Scale ruler.
- Notation sheet.
- Calculator.
- Definitions.

Definitions

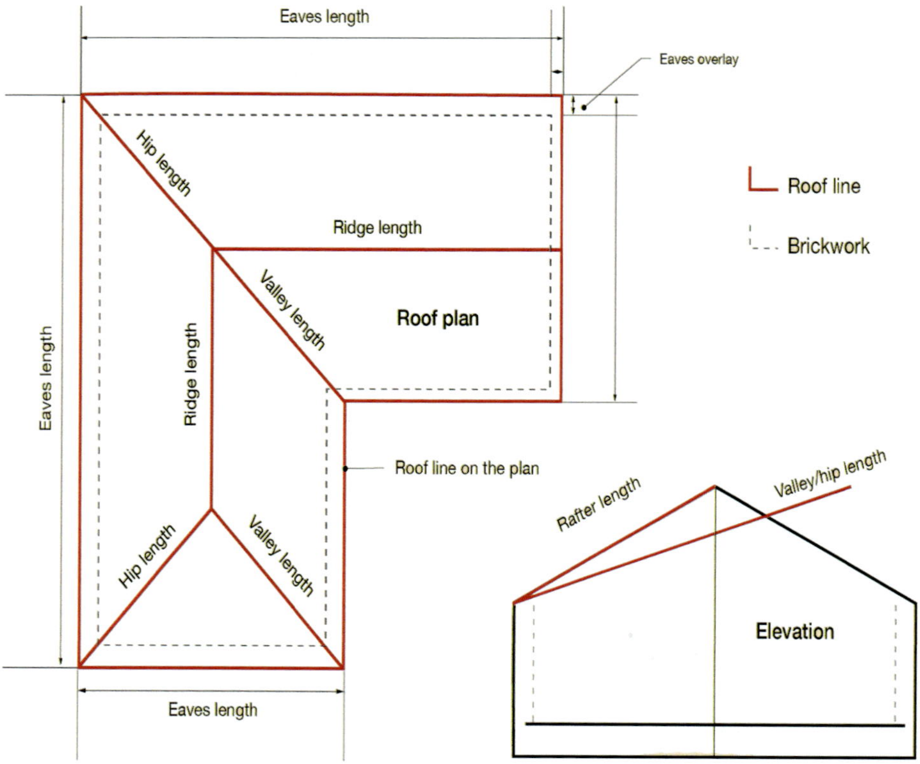

Figure 5.1 Summary of definitions diagram

Eaves length

The total length of the roof at eaves, including verge overhangs.

Roof span

The projected span of the roof on the plan, including eaves overhangs and gutter overhangs.

Rafter span

The projected span of a rafter length, equal to half the roof span for a symmetrical double-pitched roof.

Rafter length

The distance measured on top of the rafter, between the rafter apex and the edge of the rafter at the eaves or the outside of the fascia board.

Roof line on plan

The projected line of the roof perimeter on plan, including verge and eaves overhangs, and the gutter overhang.

Gutter overhang

The distance by which the tiles overhang the fascia board over the gutter. For estimating purposes with a standard 100 mm gutter, this distance is assumed at 50 mm on the plan and 60 mm on the slope. This can vary with the size of the gutter.

Roof plan area

The flat surface area on the plan is calculated from the roof line.

Roof area to slope

The actual area to be tiled.

Roof pitch

The angle between the rafter and the horizontal.

Rafter pitch = roof pitch.

Minimum roof pitch (listed in the technical data issued by the manufacturer).

Roof constant multipliers

Constant factors used in roof calculations.

Tile head-lap

The distance by which tiles overlap one another. On single-lap tiles this is the tile covering the course below it. With double-lap tiles, this is the tile covering the second course below.

Tile pitch

The angle between the tile when laid on the roof and the horizontal. The laid tile pitch is always lower than the roof pitch (approximately 5°).

Batten centres

The distance by which the battens are spaced, measured from top of batten to top of batten or centre to centre (also referred to as batten gauge).

Ridge length

The horizontal length of the roof apex.

Hip and valley length

The length of a hip or valley measured from eaves to the apex. It is advised that as the length of the hip or valley cannot be accurately measured on plan, it should always be calculated.

Estimating types of tiles and quantities

The purpose of estimating is to find out the types of tiles/fittings, and their quantities which will be used to complete the roof.

Irrespective of the method used, estimating must always be done with full tiles on both the eaves and the rafter lengths. If tiles are to be cut at a roof end to accommodate a fixed eaves length, such as between parapet walls or split gables, the perpendicular row of tiles to be cut must be estimated as full tiles. For estimating purposes, a tile to be cut is counted as a full tile and the portion cut away becomes waste.

Before calculating quantities ascertain that all areas to be tiled, including overhangs and possible overlaps of roof areas, are clearly identified.

It is always good practice to produce an additional 'notation sheet' which will help to highlight your own findings and understanding of the estimating process, and serve as a memory jogger if needed.

What drawings has the customer provided?

Often you will be doing your quantity estimate based on architectural drawings, however there will be circumstances in which you will survey buildings yourself. These could be a reroofing project where you do measurements yourself, or based on a sketch provided by someone else. The estimate will only be as good as the information you base it on, so it is good practice to check everything thoroughly before you begin.

Before taking off dimensions, examine all plans carefully for variations in pitch and any specific points needing clarification. Any doubtful points must be taken up with the customer.

Elevation drawings

Elevation drawings (example shown in Figure 5.2) are among the most common to be provided, drawn up by the architect in digital CAD and usually supplied to the estimator in PDF format.

You may also be provided with either a roof plan as shown in Figure 5.3, or you sketch your own based on the elevation view on your notation sheet. Before measuring you will need to check the scale of the drawings. For example, it may be at the scale of 1:100 when printed to A3. Often PDF readers have a measuring tool, so you take measurements on your computer or you may wish to print them out.

It is important to get the scale right as this will have a detrimental effect on your measurements if this is wrong. Often you can double check a shown measurement or a scale bar on the drawing using your scale ruler or measuring tool on your PDF reader to ensure accuracy. It is useful to remember that many doors are a standard height and this can be a good quick check on the scale ratio.

Supplied drawings might come with a floor plan, but a roofing estimator can still make use of this in conjunction with elevations to cross check dimensions, particularly if the measurement you wanted to take is obscured. You will need to remember that any brickwork dimensions will not have the roof overhang.

Once you have taken all your measurements, you can write these all down in your notes against your sketch. If you try to work to a standardised format, this will help if you have to come back to your measurements and double check something later.

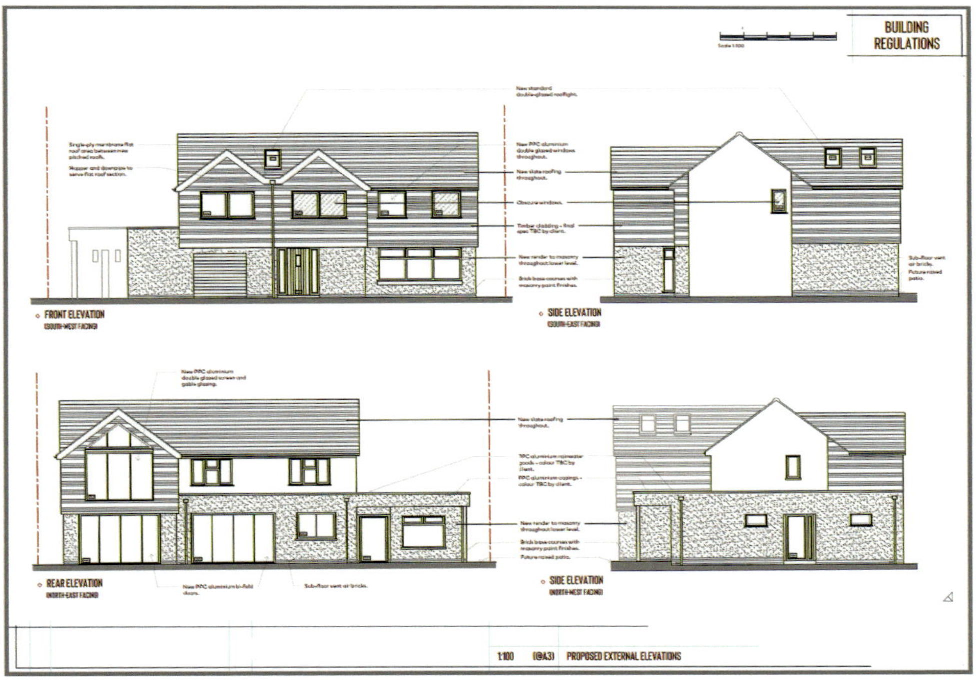

Figure 5.2 Elevation drawings

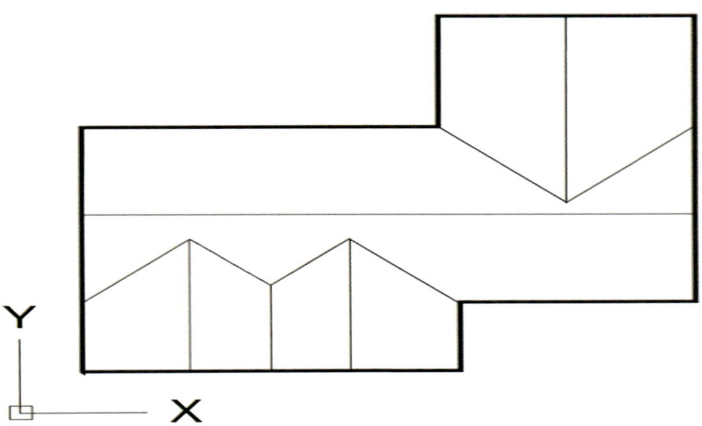

Figure 5.3 Roof plan

Determine roof pitch from drawings

It is possible to identify the roof pitch using a protractor and therefore note the pitch angle required in degrees.

In some drawing examples it is practicable to just measure the rafter length from the drawing directly, but an alternative method is useful as sometimes measuring on the pitch may not be possible. This involves taking the plan view dimensions and converting to a pitched measurement.

Converting to a pitched measurement can be a simple process. This is to convert the flat dimensions to a pitched dimension using mathematical rules with right angled triangles and trigonometry. The formulae to use is cosine of the angle θ (between the adjacent base/span and the hypotenuse) = the adjacent ÷ hypotenuse. This can also be written as hypotenuse = adjacent ÷ cos θ.

So, taking a horizontal span from eaves-to-eaves of 6.66 m. Half the span for one side of the pitched roof, at 3.33 m and pitch angle of 40° the calculation is:

Cos θ = adjacent ÷ hypotenuse
Hypotenuse = adjacent ÷ Cos 40
Hypotenuse = 3.33 ÷ Cos 40
Hypotenuse = 3.33 ÷ 0.7660
Hypotenuse = 4.34 m

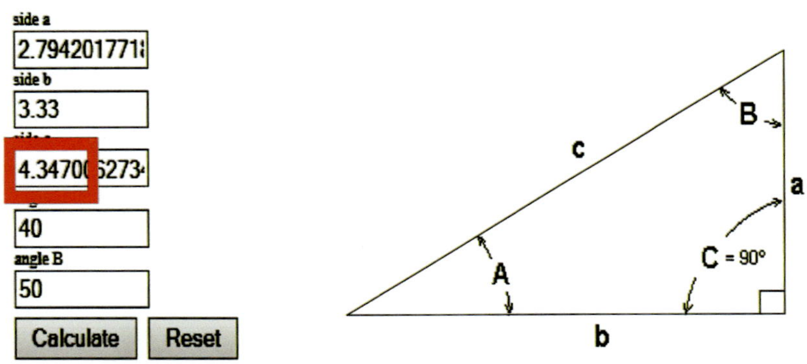

Figure 5.4 Determine the roof pitch

Alternatively, a much simpler and faster method without using trigonometry would be to use a 'rafter constant' as described next.

Rafter constants

A simple way to take your measured horizontal roof span dimension and easily convert to a pitched rafter length is to multiply against a constant. This can also be useful for hips and valleys meeting at 90° and the same pitch on either side.

Multiply the horizontal dimension by the constant for the pitch required. Table 5.1 below shows rafter and hip/valley constants at different roof pitches.

Table 5.1 Constants table

Roof pitch	Rafter constant	Hip/valley constant
12.5°	1.03	1.43
15°	1.04	1.44
17.5°	1.05	1.45
20°	1.06	1.46
22.5°	1.08	1.47
25°	1.10	1.49
27.5°	1.13	1.51
30°	1.15	1.53
32.5°	1.19	1.55
35°	1.22	1.58
37.5°	1.26	1.61
40°	1.31	1.64
42.5°	1.36	1.69
45°	1.41	1.73
47.5°	1.48	1.79
50°	1.56	1.85
52.5°	1.64	1.92
55°	1.74	2.01
57.5°	1.86	2.11
60°	2.00	2.24
65°	2.37	2.57

Calculate roof area using horizontal dimensions and constants

1. Using the appropriate scale on the scale rule (such as 1:100), measure the 'span' on the drawing, multiply by the ridge length, such as from the diagram in Figure 5.5 below.

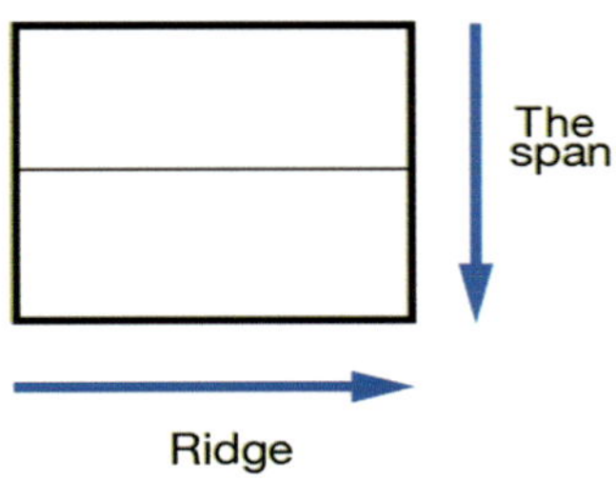

Figure 5.5 Diagram showing span & ridge

Span = 6.67 m, Ridge = 3.64 m.
Total = 24.278 m^2 (round up to 24.28 m^2)

Therefore, if the pitch is given/measured as 40° then the rafter constant is **1.31** (as taken from the constants table).

2. Rough estimate of the overall m^2 of the roof can therefore be calculated now by

multiplying the result of span x ridge x rafter constant.

Therefore 24.28 x 1.31 = 31.8m2

Calculate a rafter length using constants

If you need to know the rafter length, calculate this by multiplying the run (half of the span measurement) by the rafter constant.

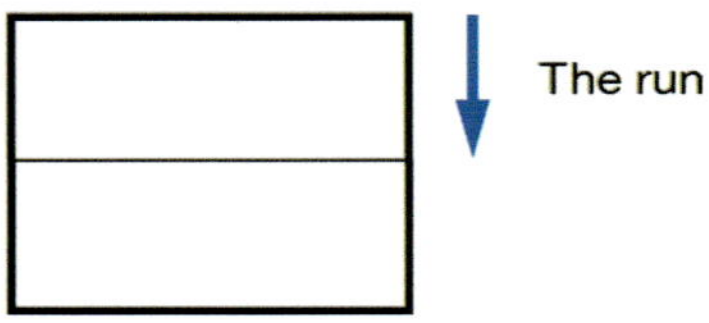

Figure 5.6 Diagram showing a run

1. With a span of 6.67 m and the run is half the span

   Run = 3.33 m, rafter constant is 1.31

Therefore, rafter length is 3.33 x 1.31= **4.34 m**

Calculate hip or valley length

If you need to know the true hip or valley length, you can calculate this in a similar way to the rafter only this time by multiplying the span measurement by the hip/valley constant.

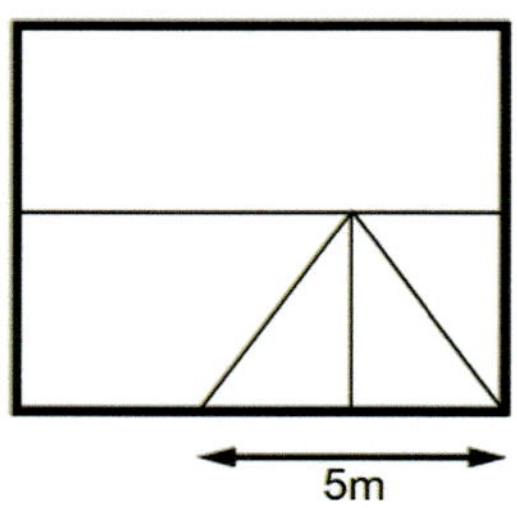

Figure 5.7 Diagram showing valley or hip

1. Span = 5 m, valley constant at 40° is 1.64

therefore, valley length is 5 x 1.64 = **8.2 m (= 4.1 m each side)**

The estimator on their notation sheet, often shows pitched dimensions noted inside a circle. More complicated roof estimates will need to be broken down into simpler parts and then added up for the grand total.

Example 1, garage

The following is an example of a simple gable to gable garage.

To begin the process, we need to identify the particular dimensions of the drawings.

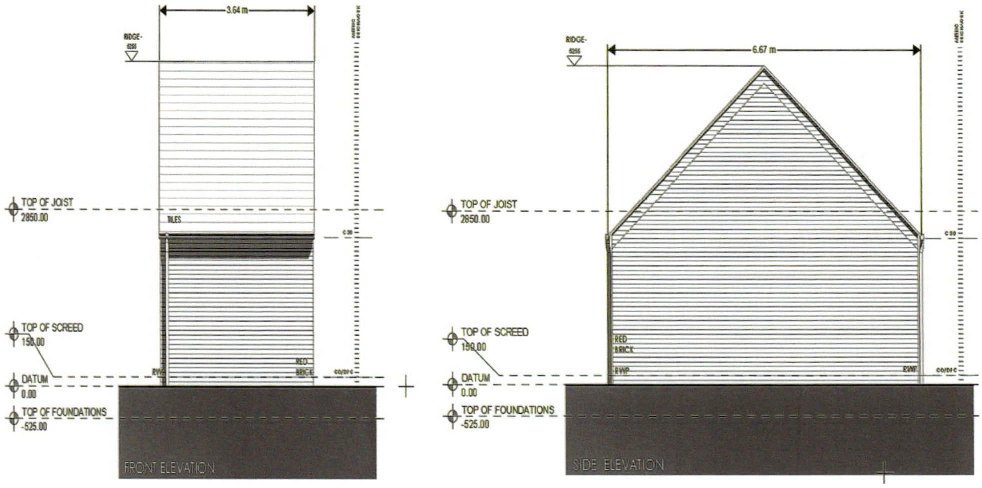

Figure 5.8 Garage dimensional drawing for roof take-off front and side

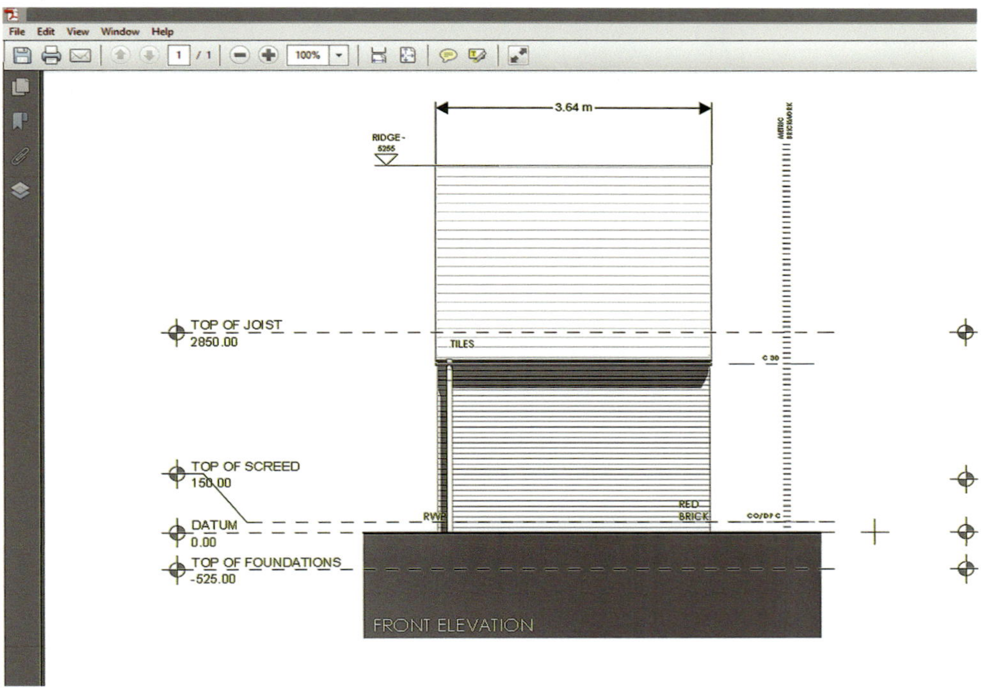

Figure 5.9 Garage dimensional drawing for roof take-off front (enlarged)

The rafter length of 4.34 can be written in a circle to denote it is a pitched dimension.

Area

Calculation is 4.34 x 3.64 x 2 = **31.59 m²**

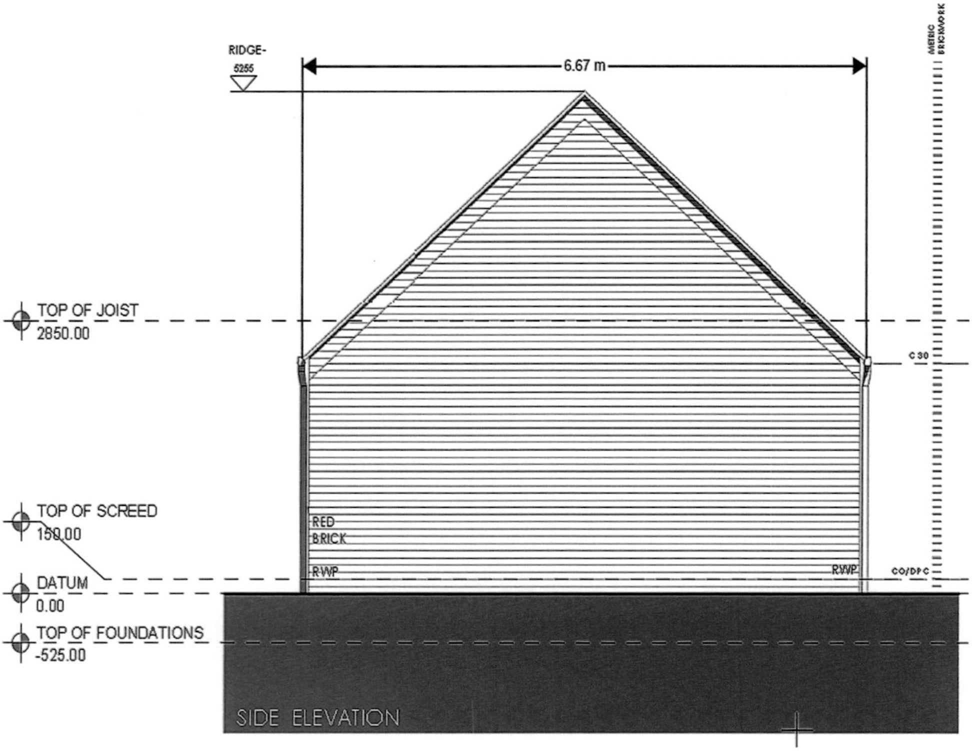

Figure 5.10 Garage dimensional drawing for roof take-off side (enlarged)

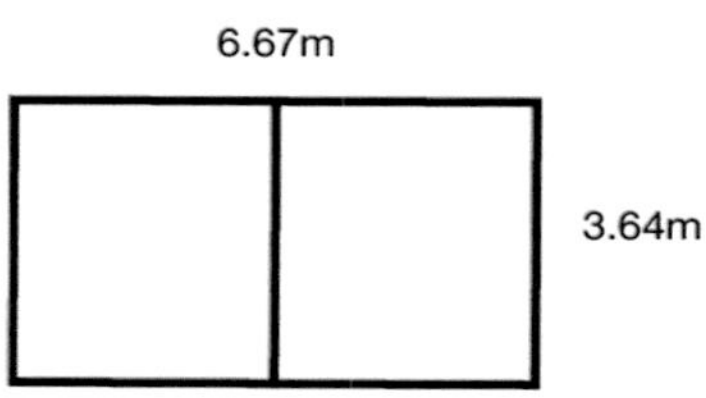

Figure 5.11 Garage ridge and span dimensions

Eaves

Calculation of total eave length is 3.64 + 3.64 = **7.28 m**

Ridge

Measurement given is **3.64 m**

RH verge

The rafter length is 4.34 therefore the right-hand verge (both sides of the roof) total is 4.34 × 2 = **8.68 m**

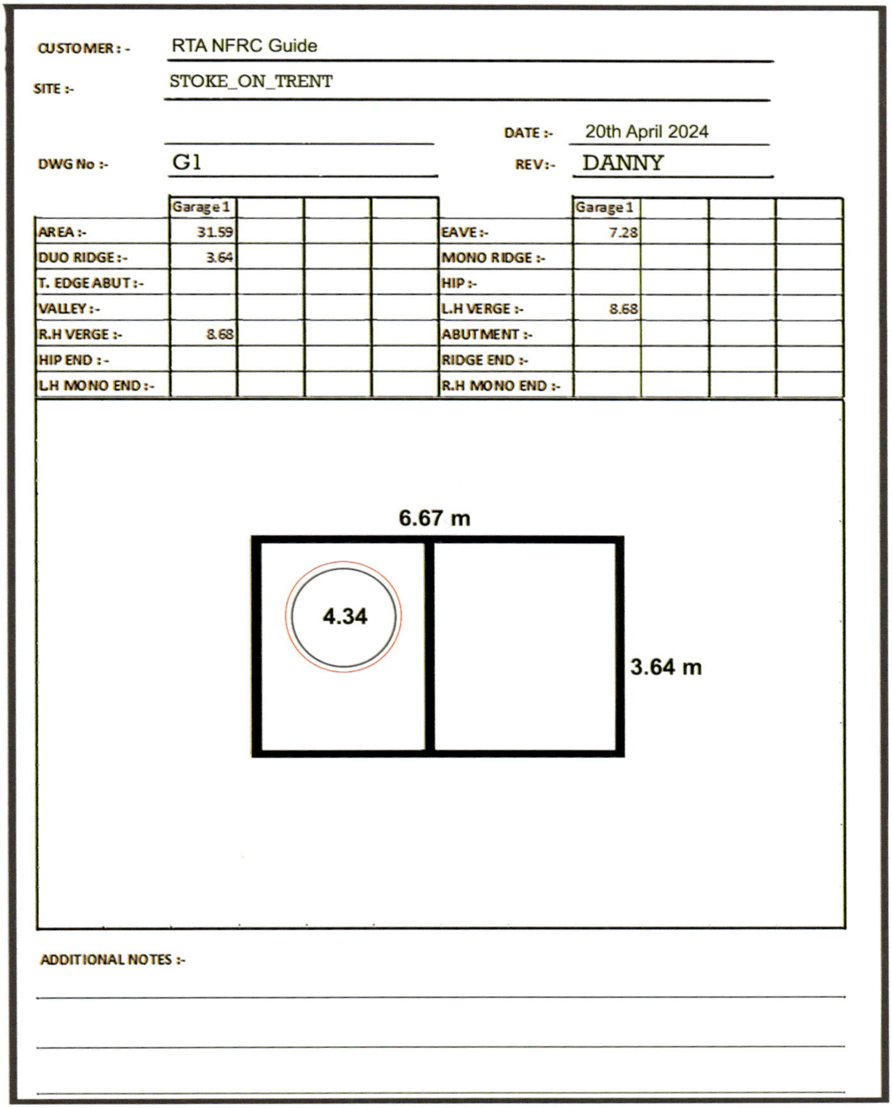

CUSTOMER :- RTA NFRC Guide

SITE :- STOKE_ON_TRENT

DATE :- 20th April 2024

DWG No :- G1

REV:- DANNY

	Garage 1					Garage 1			
AREA :-	31.59				EAVE :-	7.28			
DUO RIDGE :-	3.64				MONO RIDGE :-				
T. EDGE ABUT :-					HIP :-				
VALLEY :-					L.H VERGE :-	8.68			
R.H VERGE :-	8.68				ABUTMENT :-				
HIP END :-					RIDGE END :-				
L.H MONO END :-					R.H MONO END :-				

6.67 m

4.34

3.64 m

ADDITIONAL NOTES :-

Figure 5.12 Garage circled pitched dimensions

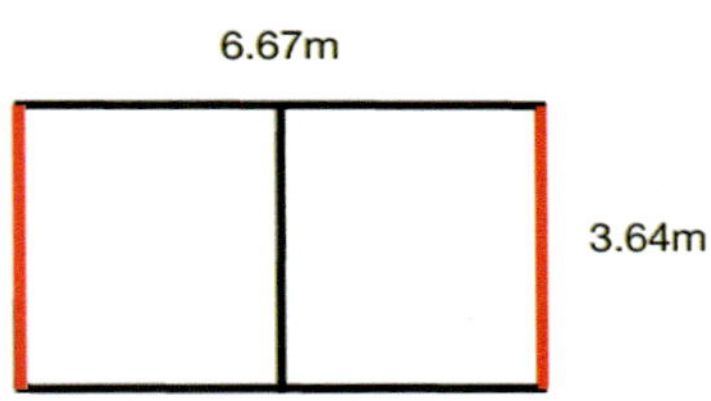

Figure 5.13 Garage diagram showing eaves

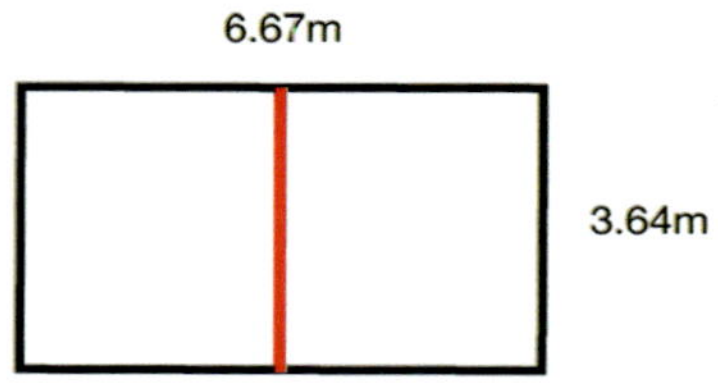

Figure 5.14 Garage diagram showing ridge

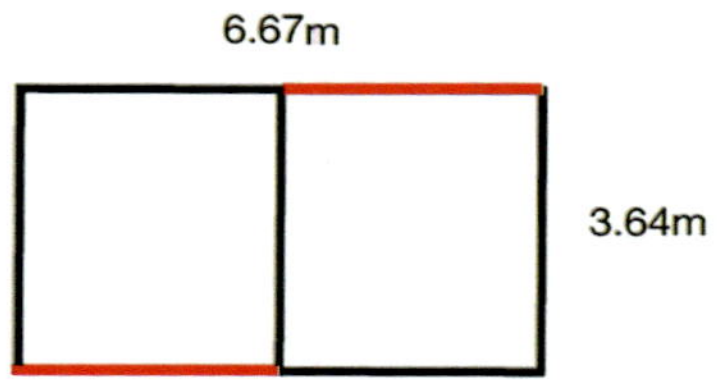

Figure 5.15 Garage diagram showing right-hand verges

LH verge

In the same way as with the calculation for the RH verge, the LH calculation is 4.34 × 2 = **8.68 m**

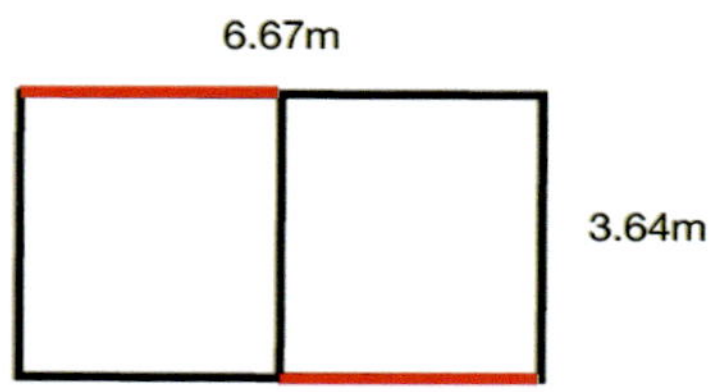

Figure 5.16 Garage diagram showing left-hand verges

Ridge end

One ridge = 2 ends

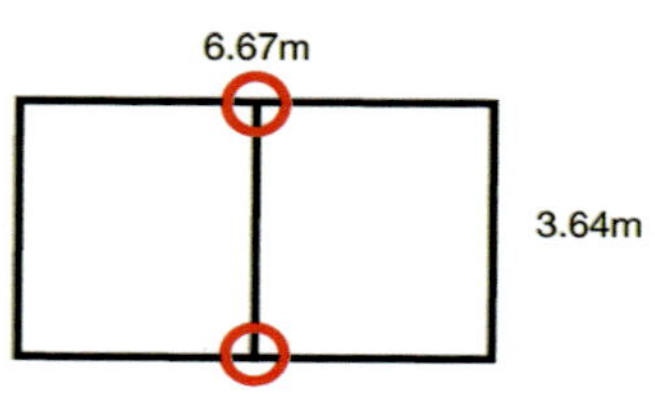

Figure 5.17 Garage diagram showing ridge ends

Converting area and linear meterage dimensions into material quantities

Each roof detail could be made up of several products, components and pack sizes required rounded up according to the particular tile manufacturer product information and pack sizes available. It is good practice to include a working wastage allowance to ensure you will not be short of materials. For example, 3–5% for tiles and fittings and 10–12% for timber battens. For the purposes of the example, a maximum fixing specification is used to demonstrate the calculation, however a bespoke site-specific fixing specification should be completed to ensure compliance in accordance with British Standards.

Interlocking tiles and fittings

Components per square metre (m^2) area

Tiles	= Area multiply by tile coverage at selected head-lap + wastage
Batten	= Area multiply by batten coverage at selected head-lap + wastage
Underlay	= Area multiply by underlay coverage based on roll size selected
Fixings	= Area multiply by fixings (subject to frequency calculation in fixing specification)

So, for example, using the garage dimensions 6.67 m span and ridge 3.64 m, plus 40° pitch, as shown in the previous diagrams above, and being covered with tiles measuring 418 mm x 330 mm, laid to a gauge of 343 mm with a head-lap of 75mm:

Tiles	@ 9.8 tiles to 1.0 m^2	= 339 no.
Batten	@ 50 mm x 25 mm	= 102 m
Underlay	@ 50 m x 1 m	= 1 roll
Fixings	@ individual tile clips	= 4 packs (100 per pack)

Ridge components per linear metre (lin. m)

Ridge tile	= lin. m multiply by length of ridge tile
Dry ridge system	= lin. m multiply by system (e.g. 5 m pack)
Ridge batten	= lin. m
Ridge endridge end cap	= each as required

So, again using the garage dimensions shown above, using ridge tiles 457 mm length and ridge roll 5.0 m length:

Ridge Tiles	= 8 no.
Dry ridge system	= 1 pack
Batten 50 m x 25 m	= 4 m
Ridge end ridge cap	= 2 no.

Eave components per linear metre (lin. m)

Eaves vent pack	= lin. m multiply by pack coverage
Eave stainless steel fixings	= lin. m multiply by clips = round up to pack quantity

So, on the garage, with standard tiles used along the eaves, eave dry vent each being 6 m length:

Eaves pack @ 10 mm eave pack = 2 packs
Eave s/s fixings @ individual eave clips = 1 pack

RH and LH verge components per linear metre (lin. m)

(**Note**: this example is using dry verges rather than tiles being cement bonded on to 150 mm wide fibre reinforced strips.)

Verge units = lin. m/units required for head-lap
Starter units = number of starters required
Verge fixings = refer to manufacturer requirements/pack quantities

So, to complete the garage roof at a verge gauge of 343 mm per verge unit.

Verge units @ RH+LH total = 26 each hand = 52 total
Starter units @ end stop unit = 2 each hand = 4 total
Verge fixings @ Drive Screw Nail = 1 pack total

You could then list your product to submit an order to your supplier as follows:

Plain tiles

Components per square metre (m^2) area

Tiles = Area multiply by tile coverage at selected head-lap + wastage
Gable tiles = Courses (rafter length divided by batten gauge) + wastage

Table 5.2 Example supplies components order list (large format tiles)

Garage Roof – example 1 **Using [insert tile name and colour]**		**Main roof**
m^2 – approx		31.59
Head-lap	75 mm	
Tiles	No	339
50 mm x 25 mm tiling batten	m	102
Underlay 50 m x 1 m roll	Each	1
Tile nails 50 mm x 3.35 mm	No	678
Tile/eave clips	No	339
Clip nails 65 mm x 3.35 mm	No	339
Eaves vent system (6 m pack)	No	2
Ridge	No	9
Ridge fixings (5 m)	Pack	1
LH verge	No	26
LH verge starter	No	2
RH verge	No	26
RH verge starter	No	2
Ridge end cap	No	2
Screw nails	No	61

Batten	= Area multiply by batten coverage at selected head-lap + wastage
Underlay	= Area multiply by underlay coverage based on roll size selected
Fixings	= Area multiply by fixings (subject to frequency calculation in fixing specification)

Note: Gable tiles (tile-and-a-half width) are required to get stretch bonded rows. These are used at one end of each unbroken course (row) length.

So, for example, using the garage dimensions 6.67 m span and ridge 3.64 m, plus 40° pitch, as shown in the previous diagrams above, and being covered with tiles measuring 265 mm x 165 mm, laid to a gauge of 100 mm with a head-lap of 65mm:

Tiles	@ 60 tiles to 1.0 m^2	= 2000 no.
Gable tiles	@ 1 per row	= 88 tiles
Batten	@ 50 mm x 25 mm	= 675 m
Underlay	@ 50 m x 1 m	= 1 roll
Fixings	@ single nails	= 4400 no.

Ridge components per linear metre (lin. m)

Ridge tile	= lin. m multiply by length of ridge tile
Dry ridge system	= lin. m multiply by system (e.g. 5 m pack)
Ridge batten	= lin. m
Eave/top tile	= lin. m per tile width (remember ridge has two sides)

So, again using the garage dimensions shown above, using half round type ridge tiles 310 mm length and ridge roll 5.0 m length:

Ridge Tiles	= 12 no.
Dry ridge system	= 1 pack
Batten 50 m x 25 m	= 4 m
Eave/top tile	= 46 no.

Eave components per linear metre (lin. m)

Eaves vent pack	= lin. m multiply by pack coverage
Eave tile	= lin. m divide by tile width

So, on the garage, with eave/top tiles used along the eaves, and eaves dry vent each being 6 m length:

Eaves vent pack @ 10 mm eave pack	= 2 packs
Eave tiles	= 46 no.

RH and LH verge components per linear metre (lin. m)

(**Note**: this example is using dry verges which each cover two tiles laid at 100 mm gauge.)

Verge units	= lin. m of verge divided by 200mm
Starter units	= no of starters required
Verge fixings	= refer to manufacturer requirements/pack quantities
Ridge end	= ridge end cap (each as required)

So, to complete the garage roof at a verge gauge of 200 mm per dry verge unit.

Verge units @ RH+LH total	= 44 each hand	= 88 total	
Starter units	@ end stop unit	= 2 each hand	= 4 total
Verge fixings	@ screw nail	= 1 pack total	
Ridge end ridge cap	@ each end	= 2 no.	

Product list summary to submit as order to the supplier: Note: Tile nails are often ordered in kg. Please check the nail weight and approximate count with the nail supplier.

Example 2, house

Add together all eaves' lengths, ridge, and total up on your notes. Calculate roof area and use the measured dimensions to work out the total linear metres of verge and valley lengths. In this example the roof windows have been ignored, but their area could be calculated and deduct from the total.

Also note that any details such as lead saddles at the ridge/valley roof junction would need to be quantified and priced. In this example, there is also a horizontal box gutter between the front pair of gables, which may be formed in lead, GRP or other specialist material as required.

Interlocking tile and fittings

Area components per square metre (m²) area

Tiles = Area multiply by tile coverage at selected head-lap + wastage

Batten = Area multiply by batten coverage at selected head-lap + wastage

Table 5.3 Example of supplies components order list (plain tiles)

Garage Roof – example 1 [insert tile name and colour]		**Main roof**
m^2 approx		31.59
Head-lap	65 mm	
Tiles	No	2000
Tile and half	No	88
50 mm x 25 mm tiling batten	m	675
Underlay 50 m x 1 m roll	Each	1
Tile nails 50 mm x 3.35 mm	No	4400
Eave/top tiles	No	92
Eaves vent system (6 m pack)	No	2
Half round ridge tiles	No	7
Dry ridge roll (5 m)	Pack	1
LH dry verge	No	44
LH verge starter	No	2
RH dry verge	No	44
RH verge starter	No	2
Ridge end cap	No	2
Screw nails	No	1 pack

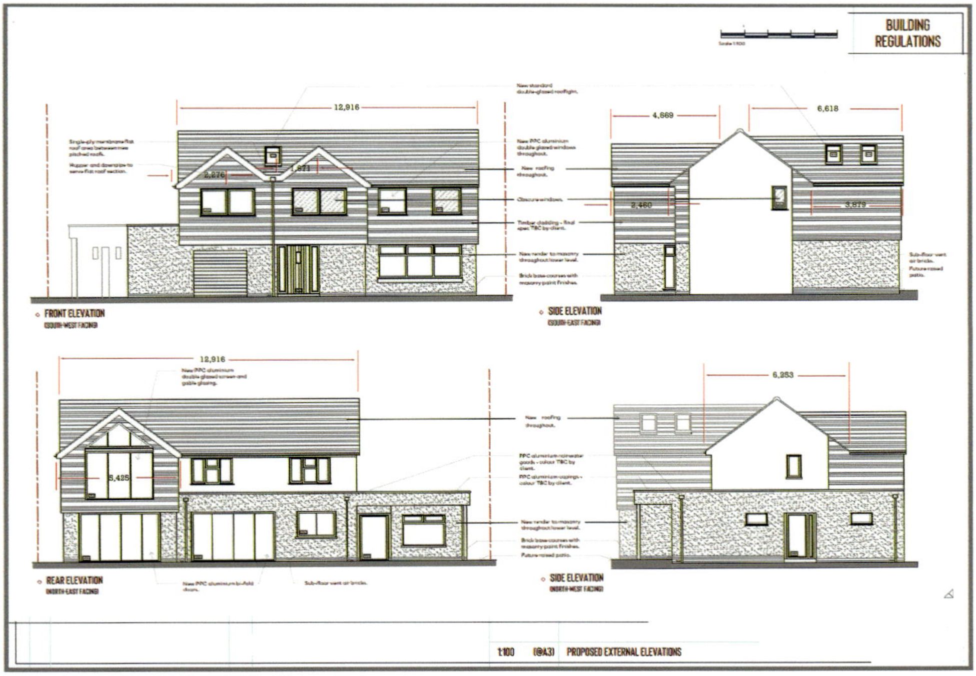

Figure 5.18 Measure your drawings

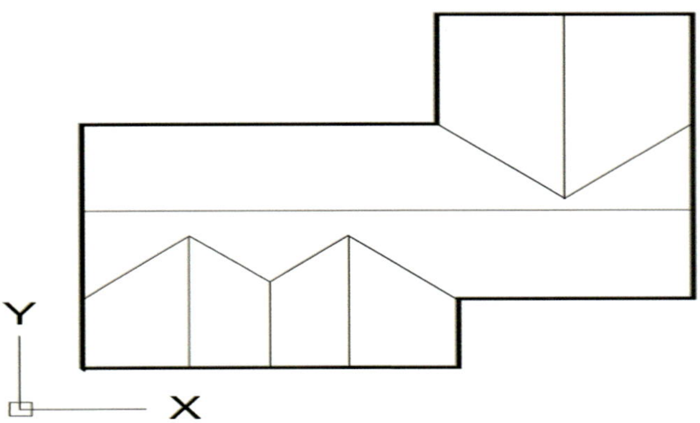

Figure 5.19 Plan view diagram

Underlay = Area multiply by underlay coverage based on roll size selected

Fixings = Area multiply by fixings (subject to frequency calculation in fixing specification)

So, for example, being covered with tiles measuring 418 mm x 330 mm, laid to a gauge of 318 mm with a head-lap of 100 mm:

Tiles @ 418 mm x 330 mm = 1593 no.

Batten @ 50 mm x 25 mm = 511 m

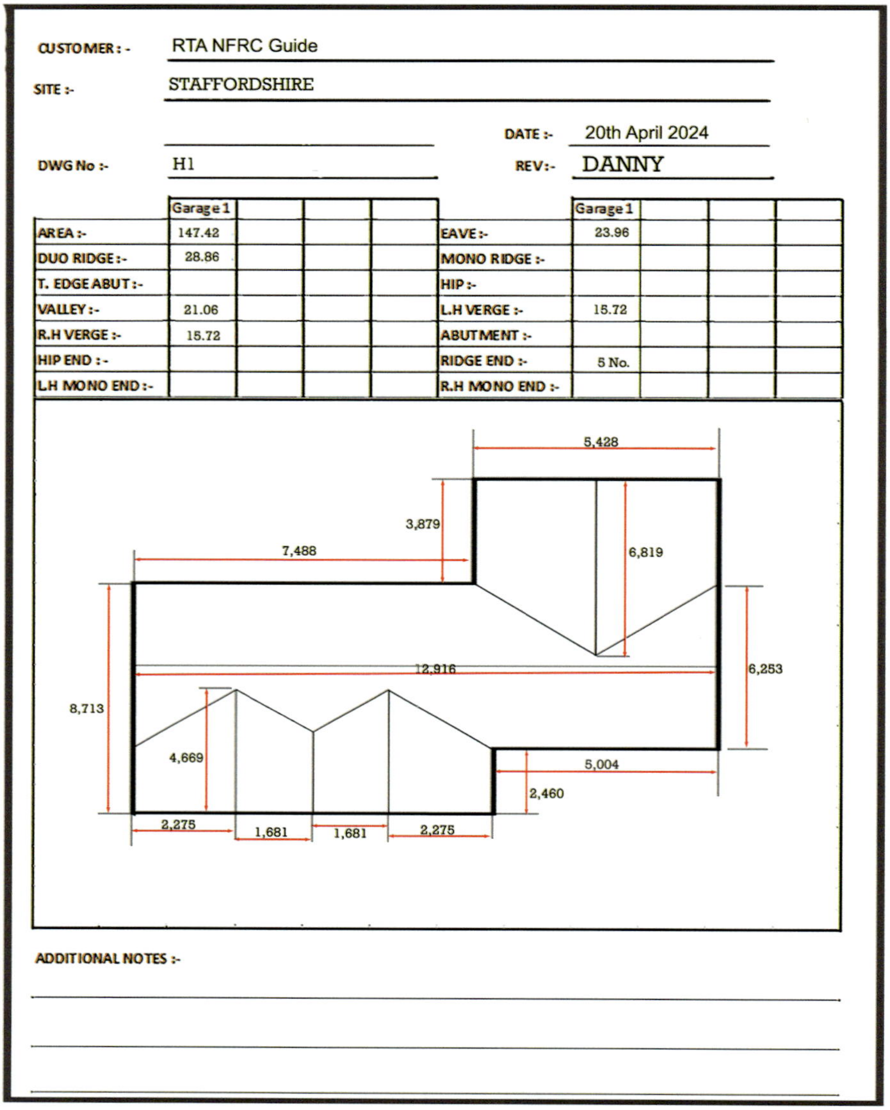

CUSTOMER :- RTA NFRC Guide

SITE :- STAFFORDSHIRE

DATE :- 20th April 2024

DWG No :- H1 REV:- DANNY

	Garage 1					Garage 1			
AREA :-	147.42				EAVE :-	23.96			
DUO RIDGE :-	28.86				MONO RIDGE :-				
T. EDGE ABUT :-					HIP :-				
VALLEY :-	21.06				L.H VERGE :-	15.72			
R.H VERGE :-	15.72				ABUTMENT :-				
HIP END :-					RIDGE END :-	5 No.			
LH MONO END :-					R.H MONO END :-				

ADDITIONAL NOTES :-

Figure 5.20 Plan view with detailed sizes

Underlay @ 50 m x 1 m = 4 roll

Fixings @ single tile clips = 16 packs (100 per pack)

Ridge components per linear metre (lin. m)

Ridge tile = lin. m multiply by length of ridge tile

Dry ridge system = lin. m multiply by system (e.g. 5 m pack)

Ridge batten	= lin. m
Ridge end ridge end cap	= each as required

So, again using the dimensions shown above in Figure 5.18

Ridge Tile @ 457 mm length	= 67 no.
Dry ridge system @ 5 m coverage	= 6 packs
Batten @ 50 m x 25 m	= 29 m
Ridge end @ each ridge cap	= 5 no.

Eave components per linear metre (lin. m)

Eaves pack	= lin. m multiply by pack coverage
Eave clips	= lin. m multiply by clips

So, with standard tiles used along the eaves and dry eave vent each being 6 m length

Eaves pack	@ 10 mm eave pack	= 4 packs
Eave s/s fixings	@ eave clips	= 1 pack

Valley detail per linear metre (lin. m)

GRP valley trough = lin. m/length of valley

therefore

GRP dry valley @ 3 m lengths with 70 mm upstand = 8 lengths

RH verge components per lin.m

Verge units	@ lin. m/units required for head-lap
Starter units	@ number of starters required
Verge fixings	@ refer to manufacturer pack quantities

So, using dry verge system

Verge units RH	@ each	= 46 units
Starter units	@ each	= 5 no.
Verge fixings Drive Screw Nail	@ per pack	= 1 pack

LH verge components per linera metre (lin.m)

Verge units	@ lin. m/units required for head-lap
Starter units	@ number of starters required
Verge fixings	@ refer to manufacturer pack quantities

Again, using dry verge system

Verge units RH	@ each	= 46 units
Starter units	@ each	= 5 no.

Product list summary to submit as order to the supplier:

Table 5.4 Example 2 supplies components order list (large format tiles)

House Roof – example 2 [insert tile name and colour]		Main roof
m^2 – approx		148
Head-lap	75 mm	
Tile	No	1593
50 mm x 25 mm tiling batten	m	511
Underlay 50 m x 1 m roll	Each	4
Tile nails 40 mm x 3.35 mm	No	3186
Tile/eave clips	No	1593
Clip nails 65 mm x 3.35 mm	No	1593
Eaves vent system (6 m pack)	No	4
Ridge	No	67
Dry ridge (5 m)	Pack	6
Valley trough 70mm x 3 m	No	8
LH verge	No	46
LH verge starter	No	5
RH verge	No	46
RH verge starter	No	5
Ridge end cap	No	5
Screw nails	No	112

Hipped roofs

Two roofs having the same eaves and rafter lengths – and the same pitch – will have the same areas regardless of whether the roof has hip ends or gable ends, as can be seen in the sketches below.

Roof configurations having the same pitch can therefore be simplified for the purpose of estimating the main roof-tile quantities.

Additional tiles for overlapping areas, if any, and for cutting work at valleys and hips, plus a percentage of wastage and breakage, are then added to the main quantity.

Roof areas having different pitches and asymmetrical or irregular shapes must be calculated and added up separately.

Calculating hip or valley lengths

Whilst no examples are shown in the above examples for how to calculate a hip or valley length, the method of calculating length is similar to using constants such as for a rafter. Whilst similar, it is more complicated with the need to consider the length is using two angles instead of just one.

This kind of dihedral angle will still use largely the same mathematical concepts of multiplying the span by the appropriate valley/hip constant taken from column in the constants table shown in Table 5.1 above, duplicated here for ease of reference.
For example, to calculate a standard valley or hip length meeting at 90° on plan, multiply the span by the appropriate valley/hip constant.

In this case, as measured from the drawing, the roof pitch is 30°and therefore the calculation shows:

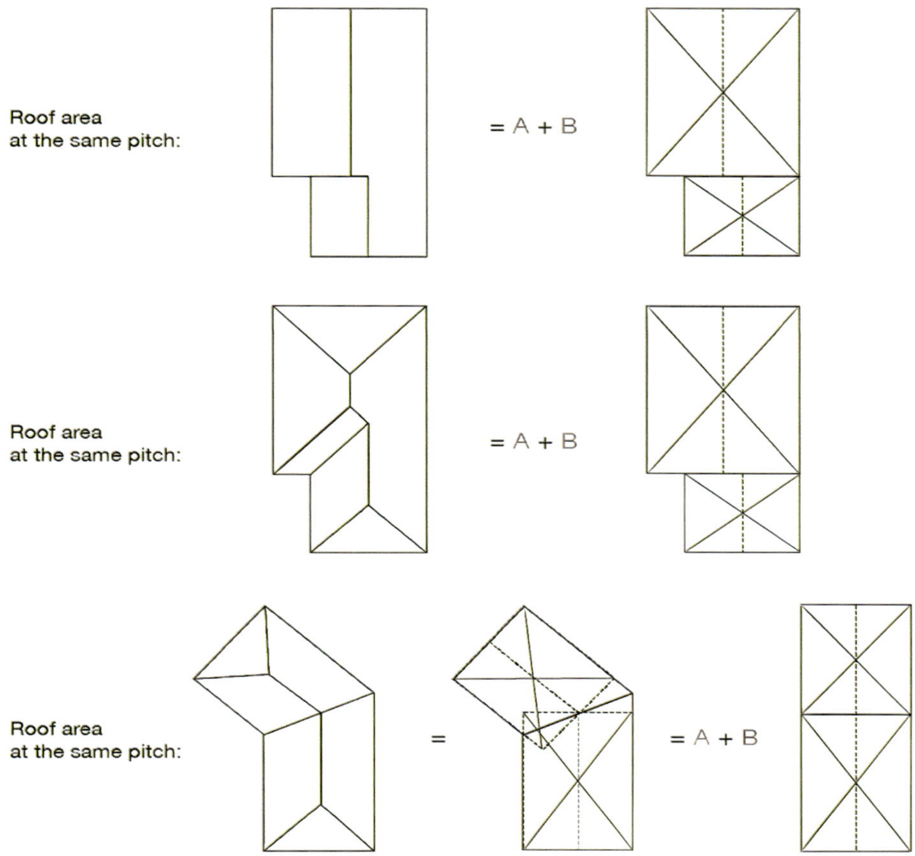

Figure 5.21 Separate different pitches and asymmetric areas

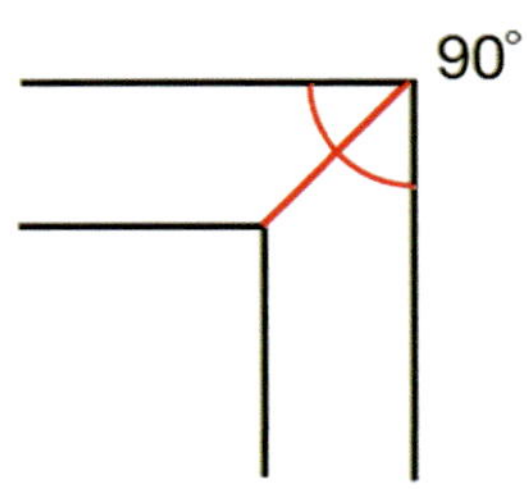

Figure 5.22 Hip or valley at 90°

Span 6 m x 1.53 constant = 9.18 m

Remember to consider that for the end hip tile it may be necessary to use a block end.

Hips or valleys for asymmetrical pitches or a splayed hip

There will be some situations where the pitch either side of a hip or valley is different and therefore the standard method of using a constant will not give accurate results. One method of simple calculation would be to use a Pythagorean theorem formula:

Copy of – Table 5.1 Constants

Constants Table		
Roof pitch	**Rafter constant**	**Hip/valley constant**
12.5°	1.03	1.43
15°	1.04	1.44
17.5°	1.05	1.45
20°	1.06	1.46
22.5°	1.08	1.47
25°	1.10	1.49
27.5°	1.13	1.51
30°	1.15	1.53
32.5°	1.19	1.55
35°	1.22	1.58
37.5°	1.26	1.61
40°	1.31	1.64
42.5°	1.36	1.69
45°	1.41	1.73
47.5°	1.48	1.79
50°	1.56	1.85
52.5°	1.64	1.92
55°	1.74	2.01
57.5°	1.86	2.11
60°	2.00	2.24
65°	2.37	2.57

$d = \sqrt{(a^2 + b^2 + c^2)}$

This is the mathematical formula used to find the longest diagonal in a cuboid. It can be applied to a roof if care is taken to use the correct dimensions.

To demonstrate the calculation, within the diagram shown in Figure 5.23, the diagonal across the base is labelled ‘e’.

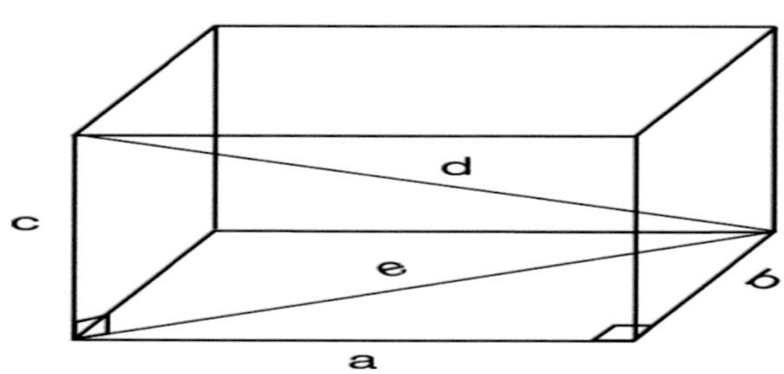

Figure 5.23 Cube with diagonals

Using Pythagoras's theorem:

$$e^2 = a^2 + b^2 \text{ and } d^2 = e^2 + c^2$$

so, $d^2 = a^2 + b^2 + c^2$

therefore, $d = \sqrt{(a^2 + b^2 + c^2)}$

Putting these formulae into context for a roof, the following uses a hip end roof, which is 25° on the main pitch and 45° on the hip end.

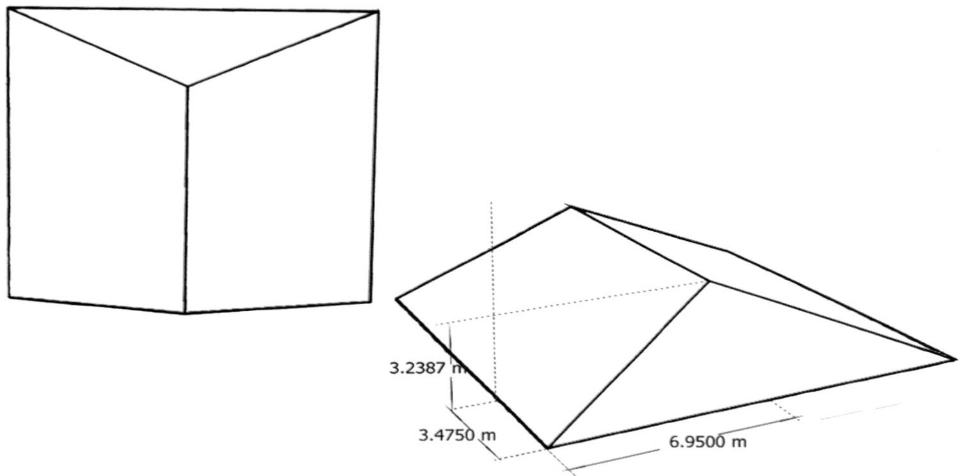

Figure 5.24 Diagram detailing a hip end roof structure

The vertical height of hip end roof structure can be taken as the same as the vertical height from the eave to the ridge at 3.23 m.

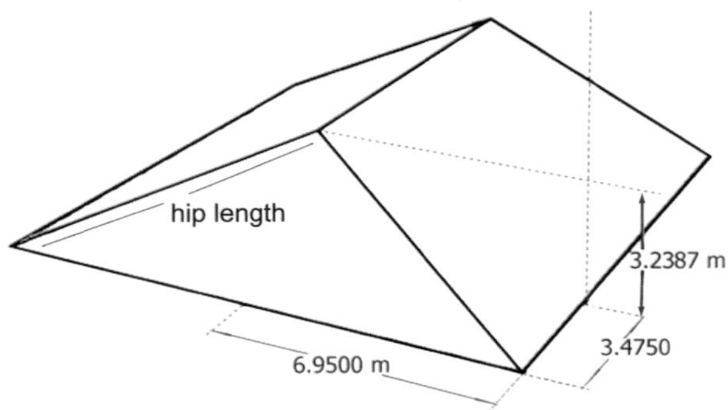

Figure 5.25 Diagram vertical height of hip end roof structure

Using the drawing measurements shown in Figures 5.24 and 5.25 above:

- the vertical height (from eave to ridge) is 3.23 m
- the horizontal half span from the corner to the middle across the end eave is 6.95 m
- along the eave from the point level with the ridge to the corner is 3.47 m

Therefore, it should be seen that the 'd' dimension is the actual hip. So the formulae calculation would be

$d = \sqrt{(a^2 + b^2 + c^2)}$

$d = \sqrt{(3.23^2 + 6.95^2 + 3.47^2)}$

$d = \sqrt{(10.43 + 48.3 + 12.04)}$

$d = \sqrt{(70.77)}$

d = **8.41** m

This formula works for any hip or valley at any combination of pitch or length assuming the three key dimensions are selected. It is important to note however, that there will often be limitations for the suitability of particular products when pitch angles are different, so it is advised that the manufacturer is contacted for advice and guidance.

Converting the measurement into quantities is a simple process of dividing it by the hip tile length to get the number of hip tiles required, e.g. 5 m hip length, 457 mm hip tile requires 11 hip tiles (rounded up). For bonnet or arris hips, you will use one hip tile per row/course of tiling.

5.3 Blending the tiles

Whilst roof-tile manufacturers cannot guarantee totally uniform tile colour or appearance, the aesthetics of a roof can be as important as the roof performance. Achieving the best visual effect from concrete and clay roof tiles requires cooperation across the supply chain.

From ordering to installation and inspection, best practice is needed to ensure best-looking roofs. Best practice mixing and blending tiles on a roof will help to minimise any issues and achieve an aesthetically pleasing final result. For more guidance refer to the current version of BS 8000–6.

This section describes issues that can influence the appearance of the roof tiles when installed. These can be inherent in the manufacturing process, caused during transport, caused during installation, or simply natural and intrinsic to concrete and fired clay materials.

Procedures

Ordering

Ordering roof tiles correctly is the first important step in achieving a good-looking finished roof. Whether the tiles and fittings are supplied from a manufacturer direct to site, or via distributors/merchants and contractors, it is recommended that the following information is supplied when placing the order:

- Customer name.
- Site name.
- Plot number (if relevant).

- Tile type and colour.
- Tile quantity.

On-site stock management and inspection

Before roof tiling commences, shipments of tiles and fittings to be used should be checked to ensure they are correct type and colour for each plot to be roofed. Any discrepancies or defective materials should be reported to the person responsible for the project as appropriate.

It is not best practice to use tiles and fittings from different shipments or batches on the same roof/plot, as different shipments will often have different concrete maturity (if concrete roof tiles) or be from kiln firings from different time periods (if clay roof tiles) as well as different climate exposure. Therefore, it is recommended that different shipments or batches of roof tiles and fittings are kept apart on multiple roof sites to avoid the possibility of mixing of tiles and fittings from different shipments. This guidance relates mainly to large format concrete and clay roof tiles, however, due to the natural variation in colours of handmade, sand faced and naturally coloured clay plain roof tiles, mixing in from different batches may be permissible with these types of tiles.

When concrete or clay roof tiles and fittings are received on site and Proof of Delivery documents are signed, it is important and recommended that a visual inspection is carried out by the contractor or developer as part of the delivery handover process. This brief inspection should include comparing the tile and fitting colour from each pallet within each shipment to ensure colour consistency and reporting immediately any major variances.

Mixing before and during fixing

It is an important requirement of all roof-tile manufacturers, both concrete and clay, that products are mixed from different pallets, normally a minimum of three. Good mixing when laying a roof minimises the visual effect of any slight colour variations within the production batch, or between different production batches in the shipment.

Adequate provision for mixing roof tiles from three pallets should be made available on site by the developer such as scaffold loading bays, loading areas and such.

Acceptance

It is recommended that before the scaffold is dropped for each plot that the site responsible person/building owner visually inspect the roof with the roofing contractor and ensures there are no obvious serious colour variation issues that would result in a problem selling the house.

Sometimes, efflorescence issues can manifest themselves some weeks after the scaffold is dropped depending on the weather. Further guidance on efflorescence can be found later in this chapter.

Examples of aesthetic issues with concrete tiles

Raw materials variation

Supply of raw materials can sometimes make an impact on the final appearance of pigmented concrete manufactured goods giving rise to some variation from batch to batch.

It is advisable to check materials at the outset and before installation to ensure a reasonable colour match exists.

Variegated colours

Variegated colour tiles are designed to be random and should be mixed on the roof to avoid distinct patches of shade banding.

Scuffing

Scuffing is where the surface of the tile is abraded, scratched, blemished or marked, often leaving white marks which are most visible on dark-coloured tiles. This is purely an aesthetic issue and does not impact on the technical performance of the tiles. The more tiles are transported, lifted and moved around a stock yard or site compound, unpacked and repacked, the more susceptible they become to issues like scuffing.

Light scuffing will weather away with time and tiles can be used with this level of scuffing. However, it is possible that the scuffing is heavy enough to remove paint from the surface of painted tiles.

If the paint has been scratched off the surface weathering will over time fade the colour of the rest of the tile to better blend the scuff marks in, but the scuff marks will remain.

Poor paint finish

Poorly painted tiles are a much rarer but occasional occurrence. All manufacturers have extensive quality measures in place, but on occasion, for various reasons, tiles can make their way to packaging without an adequate paint covering. It is extremely rare that an entire pallet will be poorly painted and so the odd tile should easily be identifiable when loading the roof and these can be set aside and queried with the supplier.

Efflorescence

Efflorescence shows as a white film on the roof tiles. It does not impact upon a product's performance and is purely aesthetic in nature.

Concrete products and mortar consist of sand, gravel, cement and water. Efflorescence is a naturally occurring phenomenon caused by water in the form of rain, condensation (on the reverse side of the roof tiles) or dew penetrating into the pores of concrete and then carrying calcium hydroxide, which is formed during the hydration process of the cement, to the outer surface of the roof tile. The water then evaporates, leaving a white film, bloom or streaking on the surface.

Efflorescence can also occur in mortar used for bedding common details such as ridge and hip tiles, where rainwater carries these chemicals from the ridge down the roof and it deposits on the roof tiles. Once deposited on the surface of the product, the calcium hydroxide then reacts with carbon dioxide in the atmosphere and becomes an insoluble calcium carbonate.

These white stains disappear naturally over time. Rainwater is slightly acidic, therefore long-term weathering will eventually remove efflorescence, but it is difficult to accurately predict how long this will take. It is important to remember efflorescence will not interfere with the functionality or durability of the tile. The short-term visual impact of any efflorescence that may occur can be mitigated by the effective mixing of tiles on the roof from different pallets.

Examples of aesthetic issues with clay tiles

Clay is a natural material made up largely of inorganic clay minerals (including kaolinite, illite, montmorillonite) and quartz (sand), along with smaller quantities of other components such as iron oxides, carbonates, pyrite, organic material, etc.

In the UK, there are many different clay locations and each source has its own mineralogy and chemical composition. These each fire differently, offering a range of colours post firing. Manufacturers will often blend two or more different clays to help achieve different tile colours, tensile strength and frost resistance.

Small changes in the mineralogy across clay seams or the blend, along with the position of the product in the kiln, can have an impact on the fired colour of the finished product. Over a period of time, slight changes can result in tiles that are either darker or lighter than a burn from months previously.

To overcome this clay tile manufacturers always recommend mixing of tiles from a minimum of three pallets at all times. This will help to blend tonal shade variation found within the individual firings, and to even out size variations due to different shrinkage rates that occur during the firing process. Failure to mix tiles will not affect their performance, but aesthetics may be compromised.

Other common colour problems

- Lead and other metal flashings.
- Tile and/or masonry cutting debris.
- Biological growth.
- Bird guano.

5.4 Roofing mortar

Mortar has traditionally been used in tiling (and slating) for three reasons:

- to provide resistance in holding in place bedded items, which alone or in combination do not have sufficient self-weight to resist disturbance or wind uplift
- as a gap filler which can be pointed to provide aesthetically pleasing finishing details
- to spread compressive loads applied to the roofing component

Due to wide scale mortar failures, in particular in 2009 and 2010, it was necessary to revise the guidance on which mixes are suitable for conventional roofing works. It was therefore necessary to reassess the relevance of mortar tensile strength with regards to the security of roofing materials.

Mortar in tension

Experience has shown that, historically, the mortar bedding of concrete or clay ridge and hip tiles, on concrete or clay tiles, was able to provide sufficient tensile bond strength to resist wind uplift for low-rise domestic buildings in the United Kingdom, provided the mortar bedding was not affected and weakened by differential movement of the roof structure.

However, evidence of increasing wind speeds in the UK, and the well-documented mortar failures in 2009 and 2010 (particularly in new housing), has shown that mortar should no longer always be relied upon to provide the required adhesive tensile strength. BS 5534 recommends that all mortar bedded components fixed at verge, ridge and hips are additionally mechanically secured to provide the necessary resistance to wind uplift. Baby ridge/hip tiles used on low-level roof details (3 m maximum height) such as bay windows and porches, may be bedded using mortar only, subject to their self-weight and the mortar quality being sufficient to resist wind loads:

- Special care will always be required to ensure that the act of applying the mechanical fixing does not cause the effectiveness of the mortar to be diminished.
- NHBC Standards Chapter 7.2, *Pitched roofs* states that all mortar bedded hip and ridge tiles are to be mechanically fixed.
- In certain situations, where it may not be possible or acceptable to mechanically fix hip and ridge tiles, the responsibility for their security is likely to fall solely on the person or company who has mixed and applied the mortar for the period of the warranty provided.

Mortar – materials and mix

General

The nominal thickness of both horizontal and vertical mortar joints is dictated by the coordinating size of the roof tiling units and is normally taken as 10 mm, exclusive of any key in the jointing surface of the units. Larger units of mortar may result where deeply profiled roof tiles are used, but the mass of mortar should be reduced using tile/dentil slips as packing for all gaps of 25 mm or greater.

Cement-rich mortars are non-resilient and subject to high shrinkage. Consequently, they are less able to accommodate movement and may contribute toward cracking occurring in completed bedding of roofing components.

The properties of a mortar, in both the fresh 'wet' state and in the finished 'hardened' state, are affected by the type of sand used and also by the cleanliness of the water. It is, therefore, important that the sand should be well graded and comply with BS EN 13139, and the water should be obtained from a potable supply.

Pre-mixed and ready-to-use mortars

Pre-mixed and ready-to-use mortars became commonplace on site for bricklayers from the 1980s and under certain conditions it was deemed acceptable to adapt the mortar (add additional cement) for roofing work. However, tests carried out in 2011 indicated that even under controlled conditions, adapted mortar is relatively low on tensile strength and also soft/chalky when cured. This, along with the high number of failures reported with such mortar in recent years, has raised concerns in the industry over the long-term durability of such mixes. As such, pre-mixed and ready-to-use mortars should no longer be recommended for roofing work.

Cement

In the UK, the most common cement used in mortar for roofing purposes is Portland cement to Class 42.5 which conforms to BS EN 197–1.

Sand

BS 1200 which distinguished two types of building sand, a type S (sharp concrete sand) and a type G (soft building sand) has been superseded by EN 13139 which, in Annexe A, defines sand as coarse, medium or fine and provides tables to determine each type. However, the common descriptives of sharp sand, grit sand and soft building sand remain in place when purchasing sand from suppliers.

Lime

Hydrated lime is available as a finely ground powder. It can be used to improve the working properties of the mortar.

Hydraulic lime is used for traditional lime mortars, which are still particularly used for conservation and restoration work.

Water

In general terms, water, which is of an acceptable chemical composition (potable) for drinking, whether treated for distribution through the public supply or untreated, is suitable for making mortars. Where public supply is unavailable, the water may be drawn from a natural source, but this may contain organic salts and therefore it is recommended that such sources of water be tested in accordance with BS EN 1008 which covers the sampling, testing and assessing suitability of water, including water recovered from processes in the concrete industry, as mixing water for concrete.

Admixtures (BS EN 934–3)

The standard specifies the requirements for performance and uniformity and methods of test for air entraining (plasticising) mortar admixtures used to modify one or more of the properties of building mortars. Mortar admixtures can be used in building mortars to improve specific properties of the mortar. However, care must be taken to use and adhere closely to the particular admixture manufacturer's dosage rate.

Pigments (BS EN 12878)

Care should be taken to prevent a mortar mixture imbalance by the excessive use of colour additives or pigments. Generally, one part pigment to 60 parts of mortar is an acceptable ratio for pigmented mortars. Mix ratios should be kept to the manufacturer's recommendations and batching to ensure consistent appearance, weight and volume.

Frost protection

Mortars containing air entrainers (plasticisers) have improved frost resistance (as well as providing much improved workability). The air entrainer gives rise to a vast number of evenly dispersed expansion chambers throughout the mass which in turn, protects the hardened mortar from the effects of freezing.

Mortar mix design

With regards to sharp or grit sands, one volume cement (with plasticiser as recommended) to three volumes of sand may produce a mix which is high in strength and highly durable but is normally unworkable for most roofing work. As such, a certain proportion of soft building sand may need to be added. Conversely, softer/finer sands may produce a highly workable mix but may not provide the necessary strength and long-term durability required for roofing work.

Some coarser building sands may be suitable without additional sharp sand due to their particle size distribution (see Table 5.5 below), although mixes with soft/fine building sand alone are no longer recommended.

Reasonable tensile bond strength may be obtained by observing good site practice in the areas of raw material selection and mix design. Reduced and/or poor bond strength may certainly result if there is inadequate cement content, excessive air entrainment or unspecified admixtures used.

Proprietary roofing mortars supplied dry in bags and/or silos are a recent addition to the market and should be mixed and applied to the manufacturer's recommendations.

Control on site

Materials on arrival at the site should be:

- checked they meet specification
- stored in a clean environment and protected against adverse weather and temperature
- water should be from the mains supply, or if from other sources it should be tested before use

Table 5.5 Standard roof mortar mixes

Recommended standard mortar mixes for slating and tiling (proportions by volume)	
Type	**Mix**
3:1 With blended sand	Soft sand and sharp sand mix with the sharp sand making up no less than one third of the sand content, to one part Portland cement and plasticiser in accordance with the manufacturer's instructions.
3:1 With coarse building sand	Some building sands are coarse in nature and have been found to have a size particle distribution in line with a blended mix of fine building sand and sharp sand. Where 70–90% of the sand is able to pass through a 0.5 mm sieve, this may be used as part of a 3:1 sand cement mix with plasticiser, in accordance with the manufacturer's instructions.
Proprietary roofing mortar	Proprietary mortars should be mixed and applied in line with the manufacturer's instructions. Evidence must be provided that the mix is designed specifically for roofing work and has been tested for workability, tensile strength and hardness of set.

Mixing:

- can be by hand or machine with the materials used to the specified ratios and percentages
- where admixtures are used, whether for workability, frost proofing or colouring, these should be added at the mixing stage to the manufacturer's instructions

Supply of mortar:

- individual mortar components may be supplied by the builder or brought to site by the roofing contractor
- proprietary dry mix mortar may be supplied on site by the builder or as a silo mix, but will more likely be supplied in bags, brought to site by the roofing contractor
- roofer should check any mortar supplied for their use is to the correct mix specification

There will be a limit on time for the use of mortar once mixed, which should be strictly observed.

Workmanship

All workmanship items for mortar are dealt with in BS 8000–6 to which reference should be made.

The following section may assist in an understanding of the use of mortar in various roofing details.

Eaves

Originally, lime mortar was used at the eaves as a filler to keep out draughts and vermin and to act as a 'beam filler'. Cement mortar may now be used in some circumstances to hold down eaves units and to resist local wind loads and to support the use of ladders.

Verges

Formerly, heavy tiles were laid with sufficient overhang to give significant protection to the bedding mortar. Other methods of forming verges avoided exposure of mortar or avoided its use.

Cement mortar is used for the construction of verges in combination with a supporting undercloak, providing additional weight and gap filing, tensile resistance against wind uplift and protection from the use of ladders and such like.

Ridges

The ridge is the most vulnerable part of any slate or tile assembly. Formerly, the purpose of ridging, in clay ware, stone or lead sheet, was to hold the uppermost courses of tiles in place against wind uplift and to provide a substantial terminal anchoring point for roof access. Lime mortar was used as a gap filler.

During the Victorian period, capped ridges were used extensively, with the cap providing protection to the vulnerable upright butt joint of the ridging so that no mortar would be exposed.

Hips

Generally, the use for hips is the same as at ridges. However, where hip tiles were bedded on lime mortar, a substantial hip iron was required to prevent the downward movement of the ridges.

Where double-lap hip tiles are used, mortar provides a compressive filler in the head-lap to prevent wind uplift. When cement mortar is used with hip tiles, a hip iron is required to restrain downward movement whilst the bond strength of the cement mortar fully develops.

Valleys

With double-lap tiles, mortar should not be exposed and where used as a compressive filler, should not impede the flow of rainwater in the side or head-lap (see BS 5534). Cement mortar is used with single-lap and profiled tiles to support the cut edges of the tiles at the raking cut of the valley trough. It also provides supplementary fixing, and it is relied upon to give extra weight to the component to resist wind uplift.

Where cement mortar is used with metal valley linings to support single-lap tiles, the mortar should be separated from the metal to allow differential movement to take place without damage to the metal or the mortar.

Where cement mortar is used with preformed valley troughs or valley tile units, the manufacturer's recommendations should be followed. Mortar used for bedding and pointing should be of the same colour throughout. Bedding and pointing should be carried out using the same mix, to ensure that the mortar sets as one piece. Veneers of mortar are not acceptable.

CHAPTER 6

Installation and execution

6.1 Laying the underlay

Roofing underlays and their use are described fully in Chapter 4.1 *Underlay*. Installed under the tiles and battens, they are laid with a drape which allows rain or condensate to drain away to the gutter.

The following describes common detailing:

Eaves and bottom edge

Consideration should be given to the following when laying underlay on the eaves and bottom edge of the roof:

- The underlay or its replacement should be detailed to extend over the fascia board and tilting fillet and into the gutter to allow effective rainwater drainage.
- Ponding or water traps at the eaves should be prevented.
- The underlay extending into the gutter should not significantly affect the flow of rainwater in the gutter.

Some underlay materials may degrade due to being subjected to weathering in this exposed position. It is recommended that a type 5U felt should be used for the exposed sections unless the underlay used for the remainder of the roof has a third-party assessment confirming its suitability for use at exposed eaves.

Alternatively, a proprietary eaves guard which extends under the unexposed section of the roofing underlay and provides adequate projection into the guttering may be used.

Verges

Underlay intended for use on verges should lap onto the outer skin of the brickwork by 50 mm in the case of an overhanging verge, onto a flying rafter. Where proprietary verge tiles or systems are specified, detailing should be in accordance with the manufacturer's recommendations which are relevant to UK conditions of use.

Ridges

For duo-pitched roofs, underlay from one side of the roof ridge should overlap the underlay on the other side by not less than the minimum recommended head-laps required for the rest of the roof.

DOI: 10.1201/9781003196990-6

For mono-pitched roofs, the underlay should extend over the monoridge and top fascia board by not less than 100 mm. Where proprietary ventilating ridge tiles or dry ridge systems are specified, detailing should be in accordance with the manufacturer's recommendations that are relevant to UK conditions of use.

Hips

Underlay courses should overlap at the hip line by not less than 150 mm.

Valleys

Underlay for use on valleys should be laid from side to side. Each course should lap past the centre line of the valley by not less than 300 mm. Where a continuous length of underlay is laid in the valley, each course of felt from either side should be cut to mitre at the centreline of the valley and lap onto the continuous length by not less than 300 mm.

Metal and plastic valley materials and units should not be laid directly onto underlays where there is a risk of adhesion. Such adhesion can inhibit the free drainage of any moisture, resulting in accelerated failure of the underlay. Likewise, adhesion can result in the premature failure of the valley material or units. Where premature failure of the underlay or lining material may happen, the underlay should be cut to the valley and lapped onto the liner.

Junctions

Underlay should overlay roof junctions by a minimum of 150 mm in each detail.

Abutments (side and top edges)

Underlay should be turned up the abutment by not less than 50 mm under the flashings.

Back abutment

Underlay should be detailed to lap over the material forming the back gutter by 100 mm to 150 mm, depending upon the pitch of the roof. Ponding or water traps behind the tilting fillet should be prevented by design.

6.2 Loading out

General

- Before tiling commences, check delivered products against the initial order and report any discrepancies or defective materials to the site agent or manufacturer.
- Pallet loads should be checked for batch codes to ensure consistency on large roof areas.
- Special fittings should be checked against matching tiles to ensure suitability before tiling commences.

Loading roof with tiles

- Load tiles and fittings out on the roof safely, supported by the battens to avoid slippage and distributed evenly to prevent overloading of the roof.
- Ensure stacks are positioned over the rafters.
- All tiles should be mixed from at least three different pallets whilst the roof is being loaded to evenly distribute any colour shade variation and thereby enhance the tile's appearance when laid.
- Figure 6.1 shows diagrams of an evenly loaded roof, and the stacking methods of both large and small tiles.

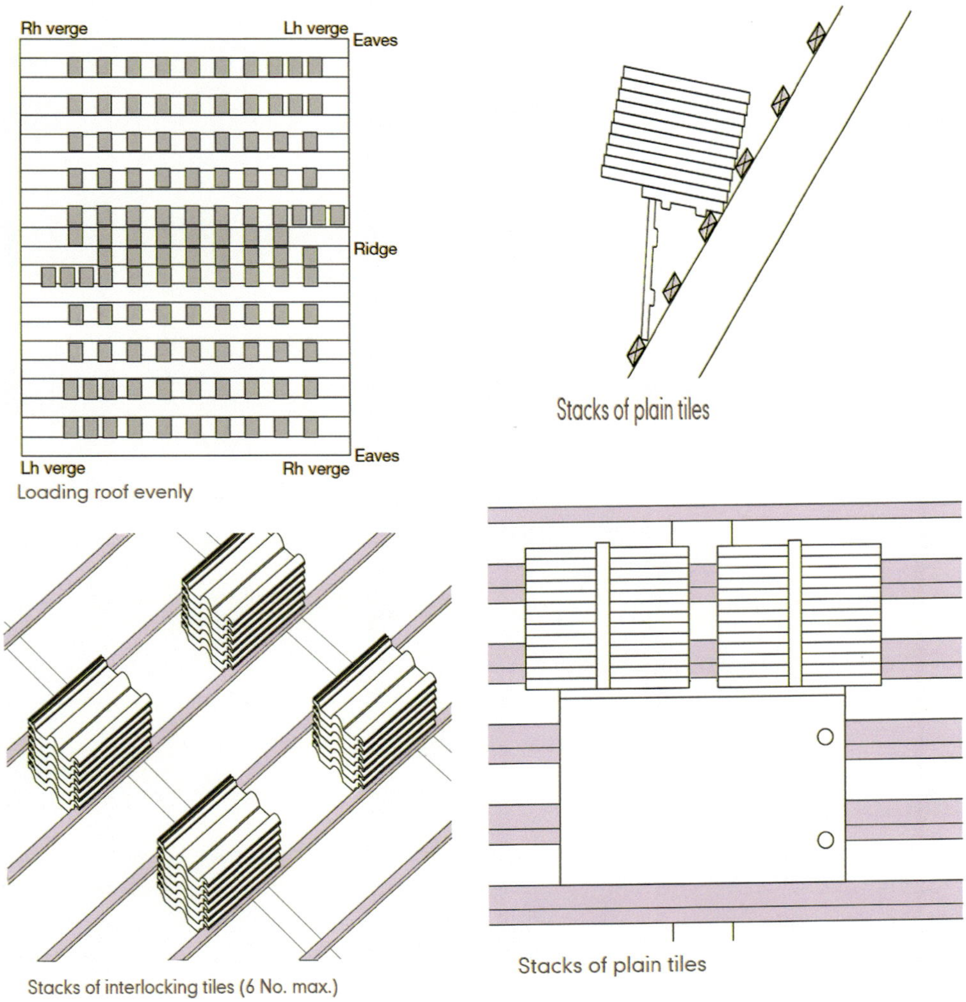

Figure 6.1 Loading out

6.3 Setting out battens and tiles

It is important that the tiler sets out the roof prior to fixing. Setting out takes place up the roof to establish the batten positions and gauges (Figure 6.2), and across the roof for tile alignment and overhangs. This will help to save time and avoid unequal overhangs at verges and expensive labour costs in cutting tiles at abutments.

The setting out of the battens needs to take into account the top and bottom of the roof and the openings through it, such as dormer windows or roof lights. The top of the roof and the bottom edge are called fixed points, and the top and bottom of each opening are also defined as fixed points.

The fixed points are used to calculate the distance between the battens. This distance is called the batten gauge, and it will be determined by the size of the tiles and the head-lap they need. It is important to note that the batten gauge may be adjusted slightly over the last few courses so long as the head-lap is not compromised.

Where two roof slopes of unequal pitch intersect at a hip or valley, set out the battens on both slopes to the lesser roof pitch.

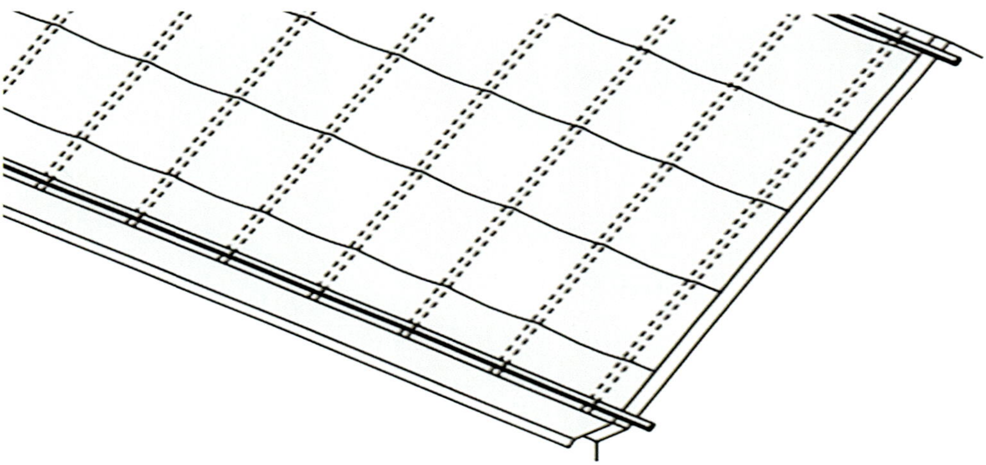

Figure 6.2 Setting out up and across the roof

Installation of underlay

Roofing underlay is installed under the tiles and is a secondary barrier against moisture in the roof space, as well as helping to provide a weathertight roofing structure. Located beneath the tiles and draped over the rafters, care must be taken during its installation to provide adequate sealed laps and to secure with the battens without compromising the water resistance.

Where an underlay overlap does not coincide with the batten, consideration should be given to either including an extra batten at the overlap or increasing the underlay lap to coincide with the next batten.

All penetrations, pipes, vents, etc. through the underlay should be suitably detailed to prevent water ingress.

Please refer to Chapter 4.1 *Underlay* which describes in detail the requirements for selection and installation.

Setting out single-lap tiles

Position of top and bottom battens

Batten gauge required must be worked out on site. Mark eaves course batten first and position using the following method:

- The eaves batten should be set to ensure that the tail of the tile extends over the fascia board and gutter by no less than 50 mm. The bottom edge of the tile should stop just short of the gutter centre (Figure 6.3).
- The end of the batten should also overhang the verge rake to the amount specified by the manufacturer depending on the verge system.
- Mark the eaves batten and measure the distance from the top edge to the outside edge of the fascia. Measure from the underside of the tile nib to the bottom tail edge to establish the hanging length of the tile. Extend a tape or rule 50 mm over the fascia board on the rake, and mark/fix the batten at this point. For example, with a tile length of 420 mm, nib depth of 20 mm, and overhang of 50 mm, then the calculation is 420–20–50 = 350 mm.
- Fix or mark the top course batten so that ridge tiles provide a minimum of 75 mm cover to the top course of tiles (Figure 6.4).

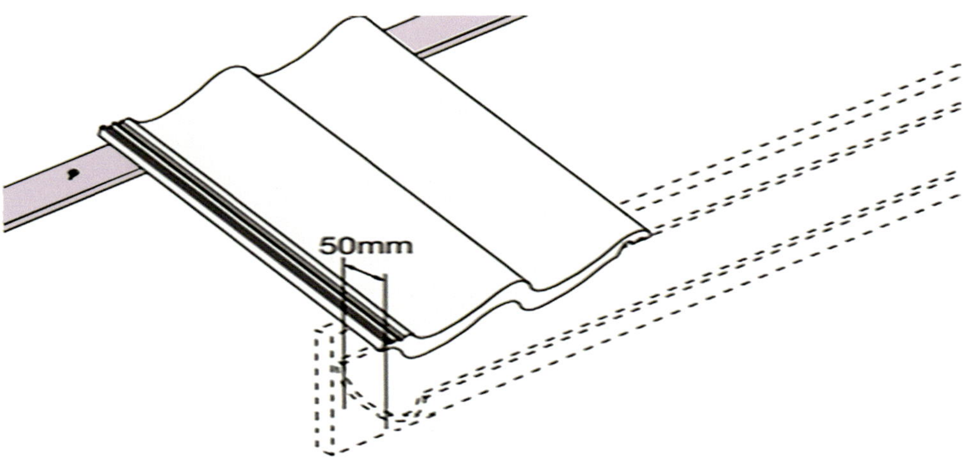

Figure 6.3 Fascia overhang

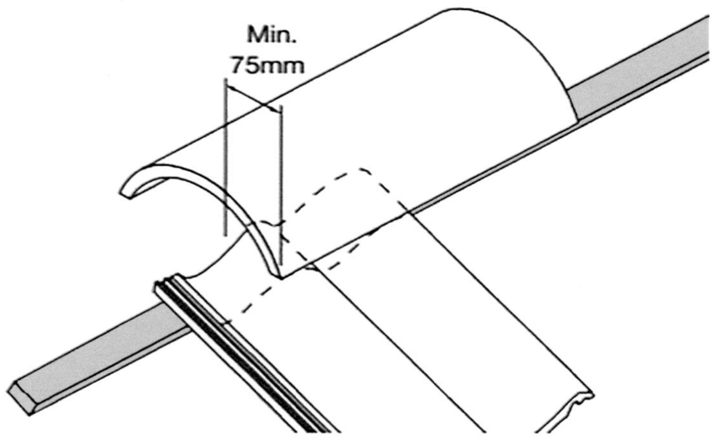

Figure 6.4 Top course batten

Calculating the batten gauge for variable/open gauge tiles

- Measure the distance between the top of the eaves batten and the top of ridge batten.
- Divide the distance by the maximum gauge of tiles being used.
- Round the figure up to give the number of courses up the slope as a whole number.
- Divide the measured distance by the number of courses to give a batten gauge.
- The practice of adjusting the gauge over the last few courses at eaves or ridge is technically acceptable, provided the maximum gauge for tile is not exceeded.
- It is important, with deeply profiled tiles, to maintain a fixed-gauge up roof to avoid a 'dog-leg' diagonal.
- If necessary, tiles should only be cut in ridge course, drilled and nailed.

Example calculation

As can be seen in Figure 6.5 below:

- distance eaves to ridge batten is 5297 mm
- maximum gauge for Marley Mendip interlocking tiles is 345 mm
- number of courses (5297 ÷ 345) = 15.35
- as we need a whole number then 15.35 rounded up gives 16
- so the batten gauge (5297 ÷ 16) = 331mm

Note

1. The above applies only to a roof pitch with no features such as dormers and/or chimneys. Batten gauges between all such fixed points should be calculated individually.
2. Where two roof slopes of varying pitch intersect, the batten gauge should be set to a lower or longer rafter pitch.

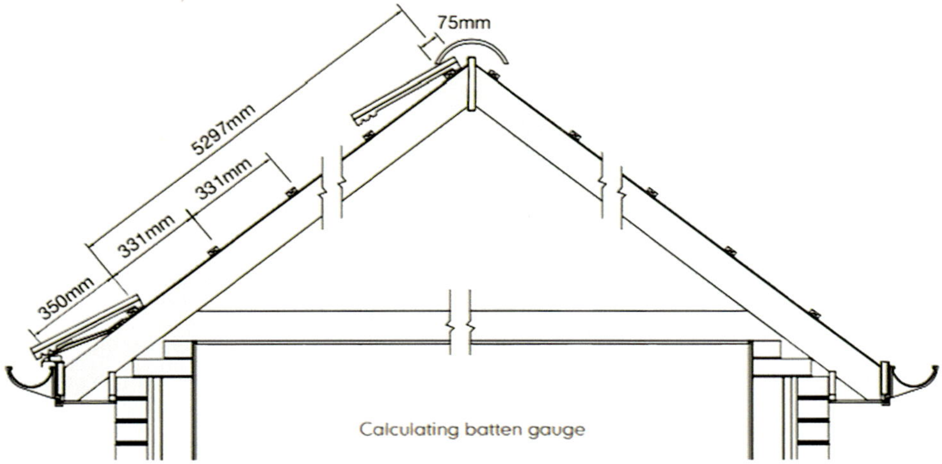

Figure 6.5 Calculation diagram

Horizontal alignment

There are several ways of achieving true horizontal alignment:

- Strike a chalk/ochre line at 90° to the perpendicular line (Figure 6.6).
- Measure two pieces of timber, each the length of the batten gauge minus the width of one batten).
- Drive nails through a length of timber, the distance of the batten gauge apart and protruding approximately 5mm. Scribe the required gauge onto the underlay.

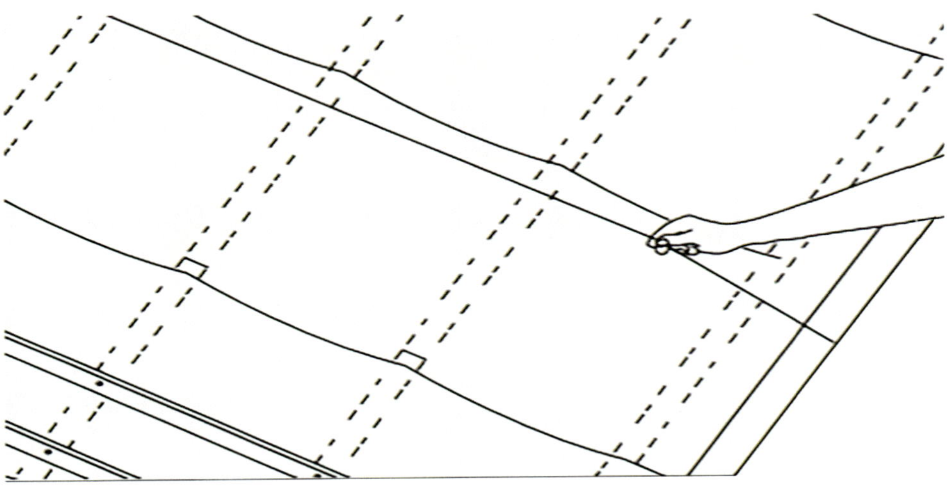

Figure 6.6 Marking the horizontal

Perpendicular alignment

- Set out the roof along the eaves starting with the correct overhang at the right-hand verge.
- Correct overhang on the left-hand verge can sometimes be achieved using full tiles by opening or closing side-lap between tiles.
- Interlocking tiles may allow a tolerance ('shunt') of approximately 3 mm in side-lock for adjustment (Figure 6.7). Set out to the average width of three tiles so that the tiles can be shunted both ways if necessary to fit the roof, allow for small deviations in tile width, and adjust for overhangs.
- Overhang at verges should not be more than 60 mm. Although many dry verge systems require 50 mm, this should always be checked with the manufacturer of the verge system being used.
- In some cases, either the right-hand or left-hand tiles will need cutting to achieve the required overhang.
- Cut tiles at verges should be at least half the width of a full tile. Easy-to-cut half tiles manufactured at the factory may be available for use at verges with flat interlocking tiles (produced in pairs for cutting on site) to enable broken bond laying.
- Strike perpendicular chalk or ochre lines over the eaves to the ridge at three-tile intervals to coincide with the left-hand edges of tiles.
- Striking or marking the batten or the use of a gauge rod the width of three tiles can be used as an alternative to actual tiles.
- A gauge rod the width of three tiles can be used as an alternative to actual tiles.

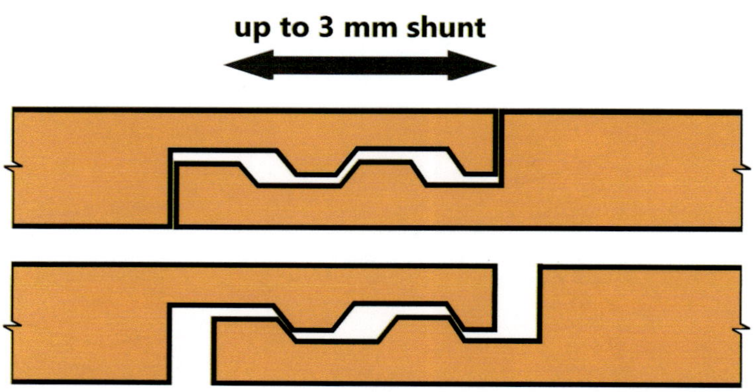

Figure 6.7 Interlocking tile tolerance

Achieving broken bond pattern

- Flat, interlocking tiles are most commonly laid half-bonded, but certain designs are laid quarter-bonded.
- Use half-width or three-quarter width tiles at verges, and cut on site.

- For quarter-bonded tiles ensure that the eaves course right-hand or left-hand verge starts with either a three-quarter width, half width or standard tile as indicated in Figure 6.8 below.
- Continue subsequent courses of tiles laid in a quarter bond, ensuring that left and right verge tiles are cut as either half tiles, three-quarter tiles or standard tiles.
- Mechanically fix all tiles by either nailing, clipping or nailing and clipping in accordance with the recommended fixing specification.

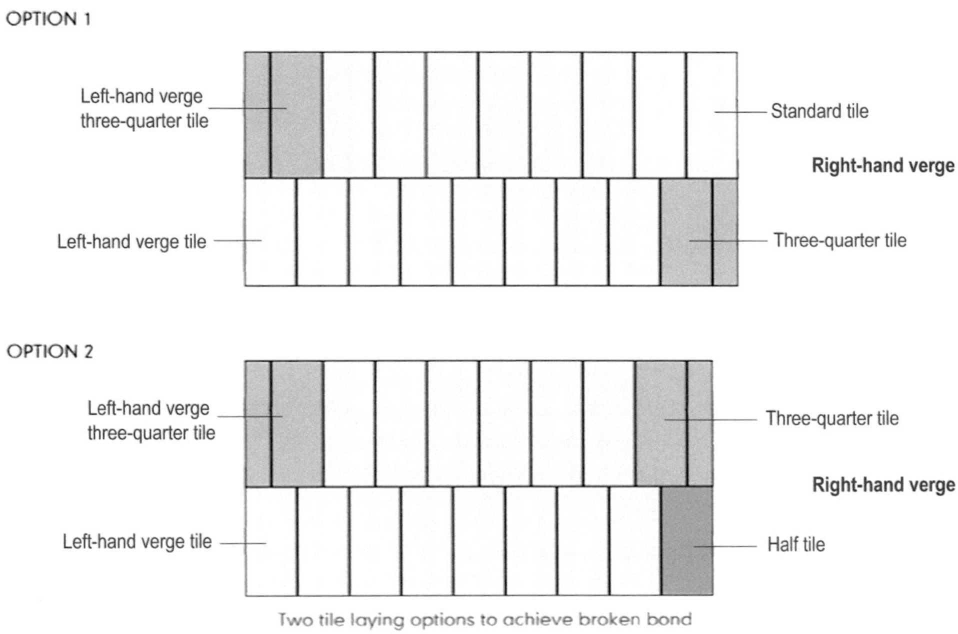

Figure 6.8 Achieving broken bond (quarter bonded)

Completion of tiling

- Load out all tiles on the roof evenly before commencing tiling.
- Work from right to left (Figure 6.9). Depending on the shape of the roof, you may need to leave out some tiles towards the left-hand verge and up the roof in columns to make use of tile battens as a ladder, enabling the upper part of the roof to be reached for fixing ridges. These tiles will subsequently need to be fixed later, in line with BS 5534.
- On a hipped roof, cut tiles so that the end tiles of each course align with the rake of the hip.

Important check points

- Never exceed the maximum gauge for tiles used at the recommended pitch.
- Avoid cutting tiles wherever possible.

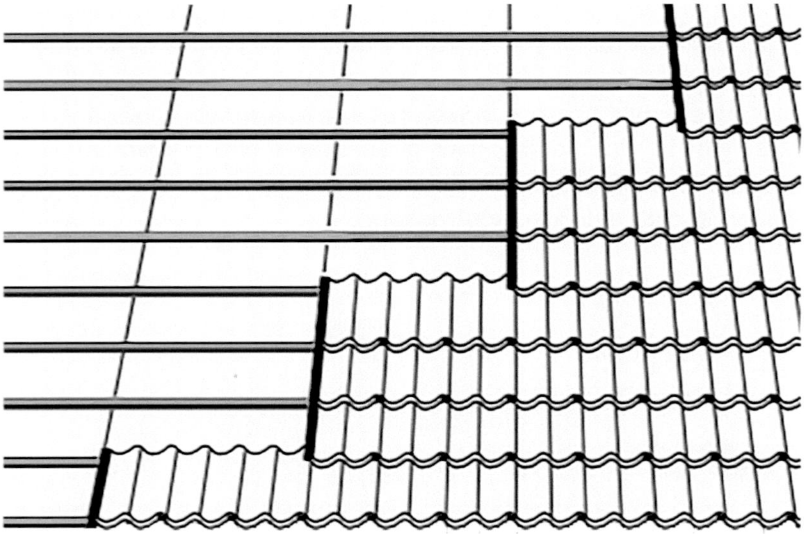

Completion of tiling

Figure 6.9 Laying the tiles to completion

- Never cut the bottom edge of a concrete tile.
- Vertical cuts should never be less than half a tile width where possible.
- On adjacent roof slopes of varying pitch, set the batten gauge to the lower roof pitch to ensure alignment at intersections, when tying in courses.
- Ensure ridge tiles provide at least 75 mm cover to top course tiles.
- Eaves tiles should extend over the fascia board 50 mm on the rake.

Setting out double-lap tiles

As with single-lap tiles, it is important that the tiler should set out the roof prior to fixing.

Setting out takes place up the roof to establish the batten positions and gauges, and across the roof for tile alignment and overhangs. This will help to save time and avoid unequal overhangs at verges and expensive labour costs in cutting tiles at abutments.

Position of top and bottom battens

The batten gauge required must be worked out on site. Fix eaves course batten first and position using the following method:

- The eaves batten should be set to ensure that the tail of the tile extends over the fascia board by no less than 50 mm on the rake (Figure 6.10).
- Position the first full course tile batten at the eaves and measure the distance from the top edge to the outside edge of the fascia. Measure from the underside of the tile nib to the bottom edge of the tile (the tail) to establish the hanging length of the tile. This distance should approximately equal the length of the tile, less nib depth and gutter overhang.
- Fix battens at the maximum gauge until about 1.5 m to 2.0 m from the ridge/apex.

- Adjust the gauge by increasing the head-lap in the last 10–12 courses only. Note that the gauge should not be increased from the eaves batten.
- Fix the first full plain tile top course batten so that the ridge tile provides a minimum of 65 mm cover (Figure 6.11). Note that with plain tiles the use of a top eave tile (which is shorter in length than a full tile) will be usual.
- Where two roof slopes of unequal pitch intersect at a hip or valley, set out the battens on both slopes to the lesser roof pitch.

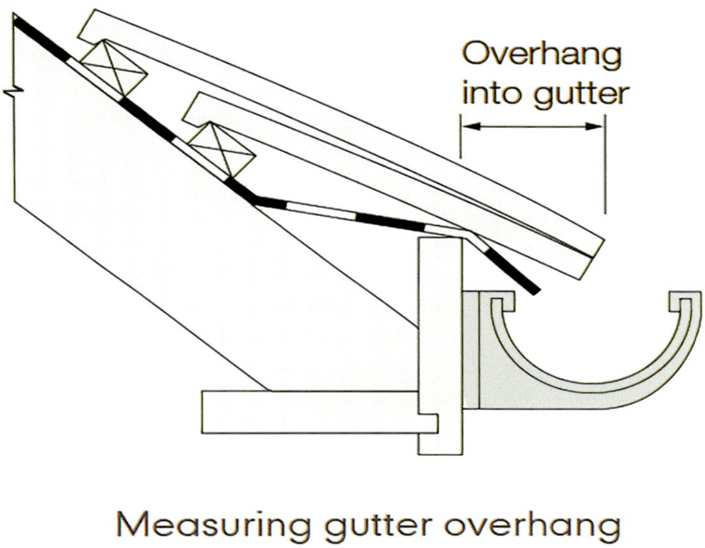

Figure 6.10 Gutter overhang

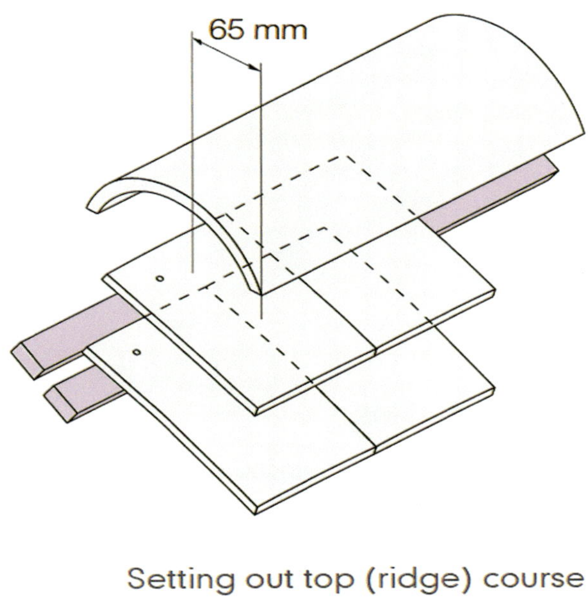

Figure 6.11 Ridge cover

Calculating the batten gauge

- Measure the distance between the top of the full tile batten at the eaves and the top of full tile batten at the ridge/apex.
- Divide the distance by the maximum gauge of tile being used.
- Gauge is determined by the length of tile less the required head-lap divided by two, for example, with plain tiles (265 mm – 65 mm) ÷ 2 = 100 mm.
- Round the figure up to give the number of courses up the slope as a whole number.
- The practice of adjusting the gauge over the last few courses at eaves or ridge is technically acceptable, provided the maximum gauge for tile is not exceeded. The better way to achieve a whole number of courses is to reduce the top batten gauges evenly by up to 5 mm for each course across the final 14 courses.

Horizontal alignment

- Strike a chalk/ochre line at 90° to the perpendicular line.
- Subsequent lines can be marked using two pieces of timber, each the length of the batten gauge minus the width of one batten.

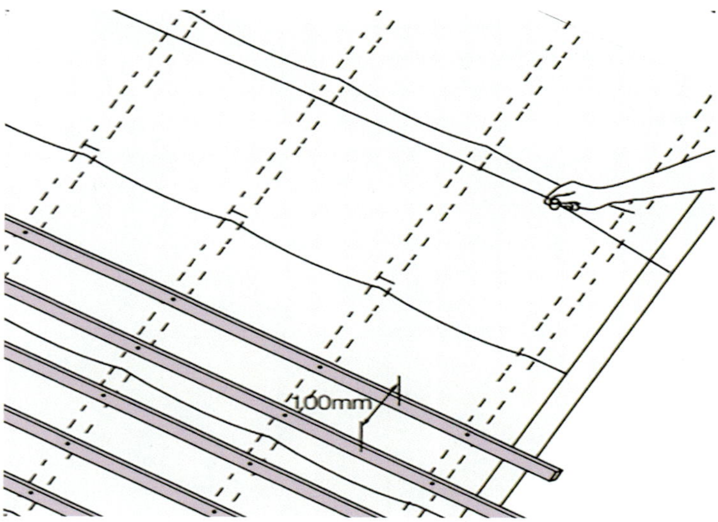

Striking chalk lines

Figure 6.12 Marking batten positions

Perpendicular alignment of tiles

- Set out the roof along the eaves starting with the correct overhang at the right-hand verge.
- Allow gaps of up to 3 mm per tile to allow for variations in tile width when setting out and tiling, to help maintain side-lap and half bond.
- Overhang at verges should not be more than 50 mm.

- Where the required overhang cannot be achieved naturally by either adjusting the undercloak (wet bedded verges only) or by opening the tiles slightly (up to 3 mm per tile) with the tiles adjacent to the verge cut to width. Many manufacturers supply tile and half-width tiles for use at verges. Cut standard tiles at verges should be avoided if possible, but if used they should be at least half the width of a full tile. (for instance, a 165 mm tile may need to be cut down to a 150 mm width).
- Strike perpendicular chalk or ochre lines over eaves to the ridge at three tile intervals to coincide with the edges of tiles.

Completion of tiling

- Load out all the tiles on the roof evenly before commencing tiling.
- Work from right to left. Depending on fixing specifications, you may leave out the third and fourth tiles from the left-hand verge and make use of tile battens as a ladder enabling the upper part of the roof to be reached for fixing ridges.
- On a hipped roof, cut tile-and-a-half tiles so that the end tiles of each course align with the rake of the hip.

Important double-lap tiles check points

- Never exceed the maximum gauge for tile used at recommended pitch.
- Avoid cutting tiles wherever possible.
- Never cut the bottom edge of a tile. The only exception is for the tile-and-a-half eave.
- Vertical cuts should never be less than half a standard tile width.
- On adjacent roof slopes of varying pitch, set batten gauge to the lower roof pitch to ensure alignment at intersections.
- Ensure ridge tiles provide a minimum 65 mm cover to top course of full-length tiles. It is even better to use top eaves tiles.
- Bottom eaves tiles should extend over the fascia board 50 mm on the rake.

6.4 Trafficking tiled roofs

Before the 2014 revision of the British Standard BS 5534, operatives installing roof tiles would leave a few tiles unfixed, then push them up so they could walk up the roof on the timber battens. This access option is now not available as the latest revision to the standard recommends that, as a minimum, all single-lap tiles should be fixed using nails, screws, and/or clipped.

It is a requirement of BS 5534 that the roof tiles, fittings and systems are appropriately fixed to resist wind loads. It is therefore crucial that the roofing contractor obtains a written fixing specification which is site-specific before work commences. Ideally, this should be done at the design stage, or certainly prior to commencement.

Perp lines

During the tile laying, there needs to be a manageable way for the operative to reach over and lay the tiles without standing on the previously installed tiles.

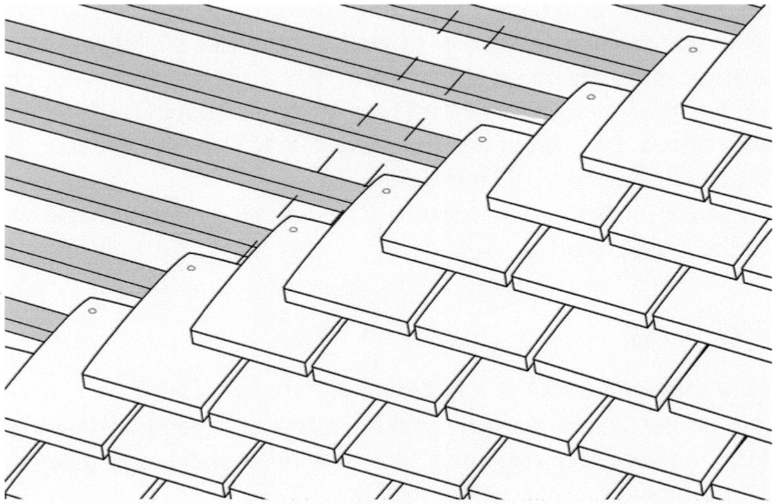

Figure 6.13 Laying plain tiles

Single-lap roof tiles should be installed from right to left with a perpendicular line (perp line) to ensure that the vertical start line of the roof covering is straight, which will enable any difference in length between the dimensions of the ridge and eaves to be accounted for during the installation of the tiles. Ideally, perp lines should be struck after every third tile so that the horizontal shunt (play) in the tile side-lock can be utilised to keep the tiles running straight which will ensure that any cutting is minimised, and where cutting is required, all the cuts will be of equal size.

Regional variations may occur where perp lines are set out in the opposite direction or the lines are struck less than every three tiles (say six for example), however, the principle remains the same.

If possible, the eaves vent system and the underlay support trays that are being used should be installed after the roof is loaded out to prevent damage caused by foot traffic during the loading process. If this is not practical (such as if the rafter roll is greater than 300 mm) then the rafter roll should be installed prior to loading followed by the underlay support tray and over-fascia vent after loading.

Position of the roofer

The finished roof covering should be installed by the roofer kneeling/standing on the timber roof battens rather than on the completed newly installed tiles. Their foot position should be where the horizontal roof batten attaches to the roof truss. At no point should the roofer stand mid-span between roof trusses.

Roof works should continue around the project working from right to left and any detailing should be completed as works progress (where reasonably practicable).

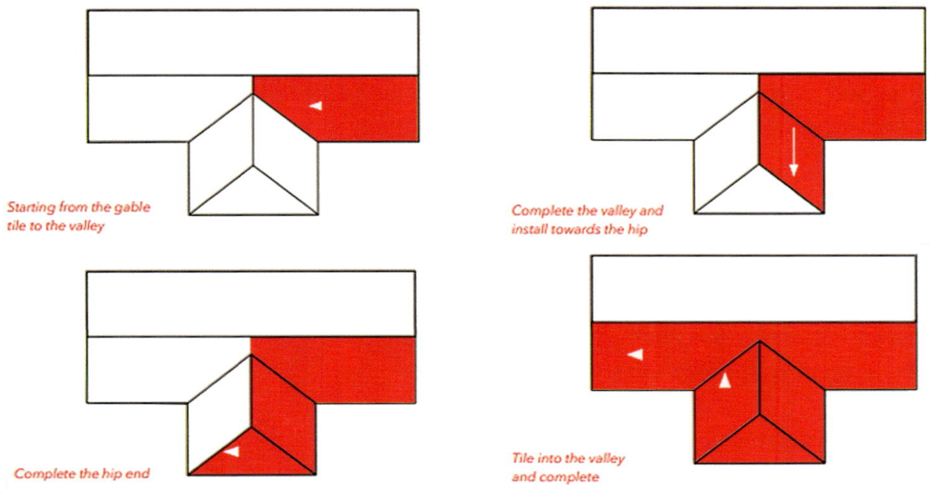

Figure 6.14 Direction of work

Roof vents

Where specified, if the position of the roof vents is known, they should be installed at the same time as the roof coverings, since installing as a retrofit at a later date would require the removal of the finished roof tiles which may cause unnecessary damage to adjacent tiles and their fixings.

Lead work

For the installation of any lead work (or equivalent) soakers and flashings, all chases into masonry or brickwork and any associated lead work required, should be cut and installed prior to the finished roof coverings being installed. This ensures that operatives are working off the timber battens with only the final dressing being undertaken once the finished roof has been completed.

Dry fix ridge/hip system

Lay the rollable weathertight membrane centrally along the ridge or hip batten (if required) and tack with a staple or underlay nail to the ridge batten.

Once the membrane has been dressed and stuck to the profile of the roof tiles, then the ridge/hip tiles should be installed as the works progress with the operative working from the leading edge of the installation off the timber battens.

Verge detailing

Verge work should be completed off a safe working platform, scaffold or ladder so that access over the completed roof coverings is not required.

If a safe working platform is erected prior to the roofing commencing, then this should take into account the distance it requires to be away from the building in order to ensure that the detailing can be completed without the need for repositioning or dismantling of the platform.

Scaffolding

Scaffolding should only be erected by qualified persons and they should follow safety guidance issued by the National Access and Scaffolding Confederation (NASC) or similar guidance provided by the manufacturers of the scaffolding system.

It is always advisable for the roofing contractor to agree, prior to any works commencing, the access and egress strategy during the installation of the roof with the main contractor or the scaffolding contractor. This will ensure that the scaffold is erected to the requirements of the tile installation contractor so that works can progress without the need to adapt the scaffold edge protection or working platform. It will also ensure that the scaffold contractor knows how to dismantle the scaffold without accessing the roof in order to mitigate the chance of damage to the finished roof.

Repair and maintenance

If at any time, access over the finished roof is required for maintenance, repair or the installation of consumables, this should be done from a kneeling position or from crawling boards or roof ladders, suitably packed with a solid foam type product (like insulation board) or other compressible material.

This will spread the load to avoid point contact on the tiles. Any access equipment utilised should be properly supported and anchored to prevent slippage or tipping.

Any cracked or damaged tiles should be replaced and secured in accordance with the specifications, using a fixing system recommended by the tile manufacturer.

Stripping old roofs

It is a fact that tiling battens deteriorate with age. They should therefore not be used as footholds unless they have been inspected by a competent person and confirmed that they are good and strong. If there is any doubt, the battens should be regarded as fragile.

When stripping softwood timber battens or sarking which may have been sourced or manufactured between 1950 and 2007, it is likely that they would have been treated with preservatives that contained hazardous substances in quantities above the hazardous threshold contained in the Environment Agency guidance of waste management WM3. These timbers should be deemed Hazardous Treated unless independent laboratory test evidence confirming otherwise is obtained. As these are hazardous waste, they will require specialist disposal contractors to remove them from site. They should always be segregated from other waste materials.

Windy conditions

It is unsafe to work or handle materials in windy conditions. The HSE recommends that tiling work should cease if the mean wind speed reaches 23 mph (gusting to 35 mph or over) and if handling rolls of underlay the limit reduces to 17 mph (gusting to 26 mph or over).

6.5 Valleys and gutters

This chapter covers recommendations for the installation and execution requirements for common gutters and valleys.

Valley with valley tiles only

Ensure that continuous support is provided for the ends of tiling battens on each side of the valley. Cover the valley with a strip of underlay not less than 600 mm wide underlapping the general underlay.

Cut adjacent tiles and tile-and-a-half tiles so that valley tiles course in and fit neatly. Note that it is not necessary to mechanically fix the valley tiles.

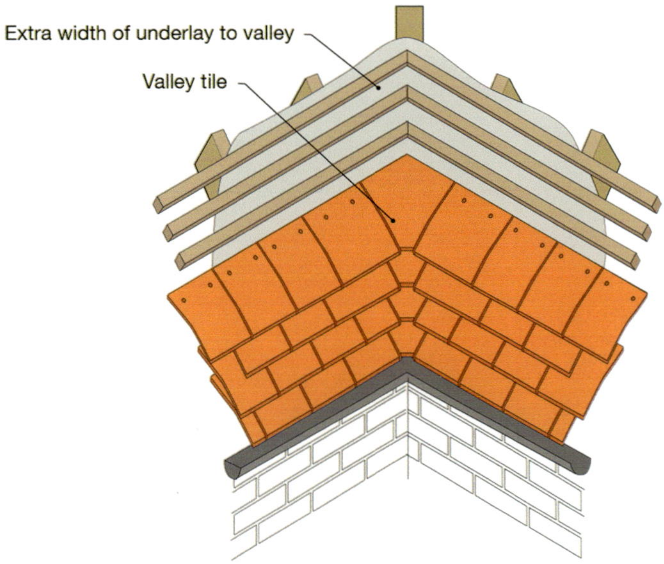

Figure 6.15 Valley tiles

Valley with tiles and soakers

Ensure that continuous support is provided for the ends of tiling battens on each side of the valley. Cover the valley with a strip of underlay not less than 600 mm wide underlapping the general underlay. Cut tile-and-a-half tiles and fix them to form a straight, weathertight, close mitred junction. Interleave mitred tiles with metal soakers, extending a minimum of 150 mm to each side of the valley. Fix soakers by turning down over heads of mitred tiles.

Valley with metal lining

Ensure that valley boards, plywood sheathing and tilting fillets provide full support for the metal valley. Cut the underlay to rake and dress over tilting fillets to lap onto the metal valley. Ensure that the underlay is not laid under metal. Cut tile-and-a-half tiles neatly and fix to form a gap a minimum of 125 mm wide centred on the valley. Either lay tiles dry or bed on mortar onto fibre cement undercloaks laid loose on each side of the valley, ensuring at least a 25 mm gap between mortar and the tilt fillet. To avoid the problems associated with open valleys, the use of purpose-made valley tiles should be encouraged wherever possible.

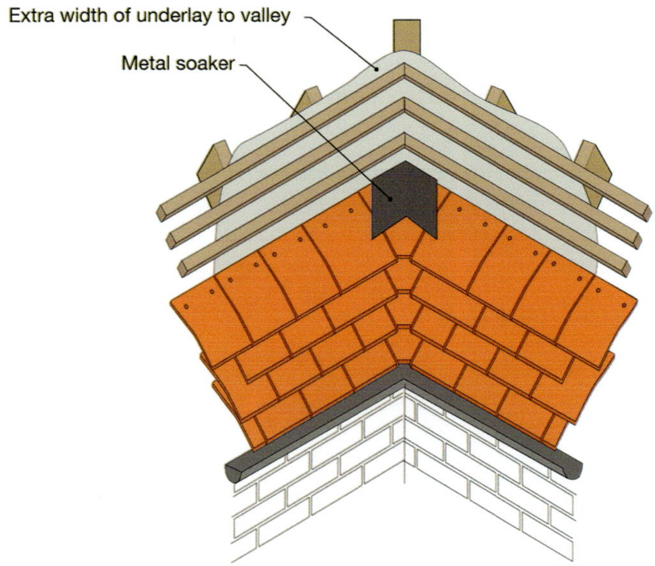

Figure 6.16 Metal soakers

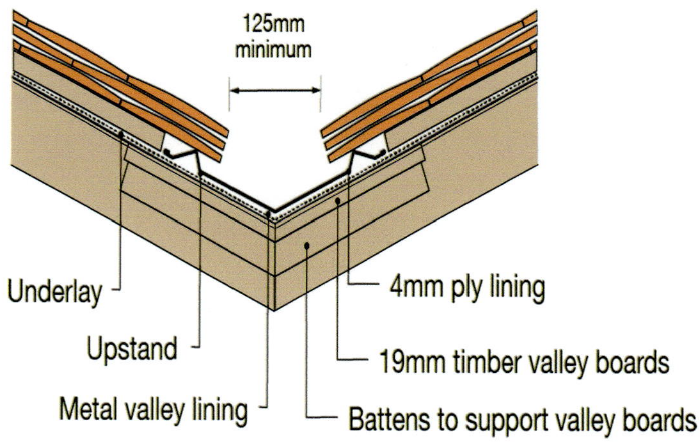

Figure 6.17 Metal lining

Valley with preformed GRP valley

Ensure that valley boards provide full support for the glass reinforced polyester (GRP) valley. Some types of GRP valleys do not require valley boards – check with the valley manufacturer. Lay the underlay as recommended by the valley manufacturer. Fit the valley as recommended by the valley manufacturer.

The valley should be secured to counter battens by at least one nail every 400 mm. Cut tile-and-a-half tiles neatly and fix them to form a gap a minimum of 125 mm wide, centred on the valley. Either lay dry or bed on mortar onto the GRP valley. To avoid the problems associated with open valleys, the use of purpose-made valley tiles should be encouraged wherever possible.

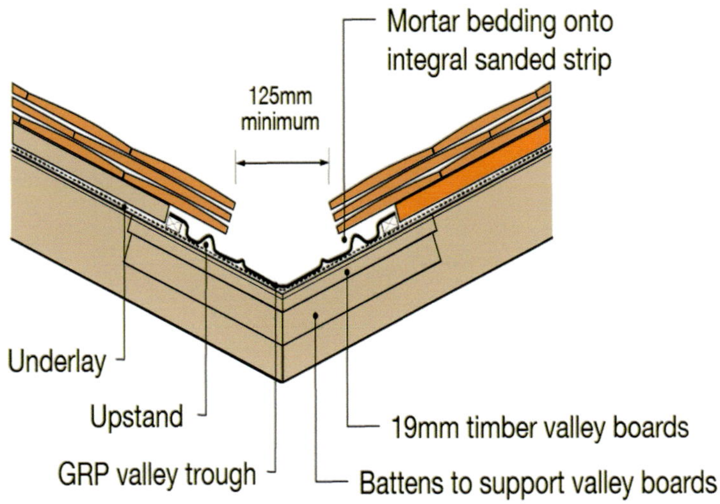

Figure 6.18 GRP valley

Box gutter eaves

Fix a continuous timber tilt batten at the eaves to provide support for metal flashing and eaves course of tiles. Cover the metal gutter lining up over the tilt batten and extend under the eave course by a minimum of 150 mm to 200 mm. Lap the roof underlay over the metal welt.

An over-fascia ventilation strip could be positioned on top of the tilt fillet to provide low-level ventilation, but airflow may be restricted, depending upon the height of the wall. Also, the designer must be satisfied that the gutter will not become blocked with debris or by freezing and cause flooding into the roof space through the ventilator. If either of these points are of concern, then ventilation tiles should be provided clear of the wall and gutter instead.

Bonding gutter with metal lining

Ensure that the gutter board is wide enough to provide full support for the ends of tile battens, tilt battens and the valley lining. Pack the void between the board and the top of the party wall with quilt insulation. Turn the underlay over the timber tilt battens on either side of the bonding gutter. Use tiles and tile-and-a-half tiles to form a gap no greater than 15 mm between plain tiles and adjacent roof covering.

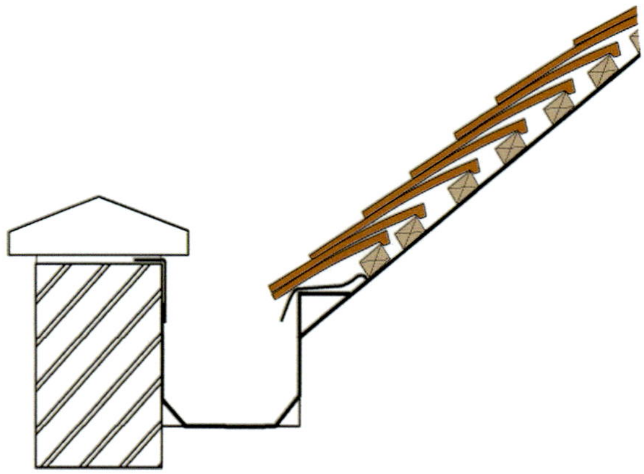

Figure 6.19 Box gutter eaves

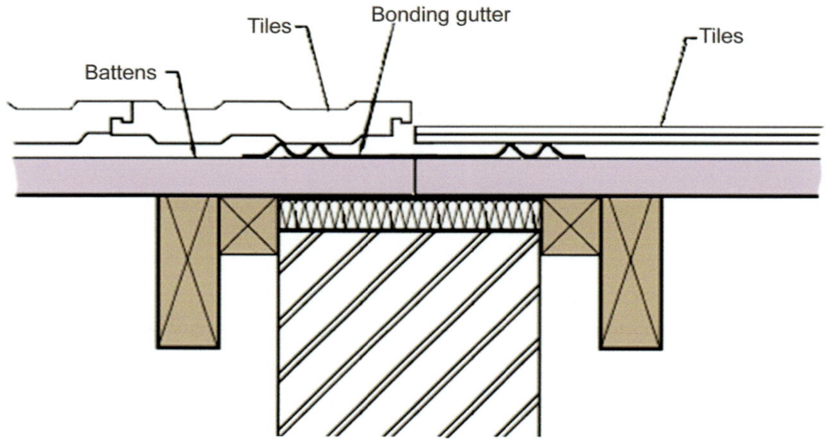

Figure 6.20 Bonding gutter with metal lining

6.6 Flashings and abutments

Pipe flashing

Figure 6.21 shows a standard metal flashing for use with a 100 mm diameter pipe. The metal flashing should extend a minimum of 150 mm below and to each side of the pipe and at least 100 mm under the tiles above the pipe, ending in a welt. The flashing should extend at least 150 mm up the pipe, measured at the back of the pipe, and can be turned into the pipe. Alternatively, a pipe collar can be fitted to weather the joint between the flashing and the pipe.

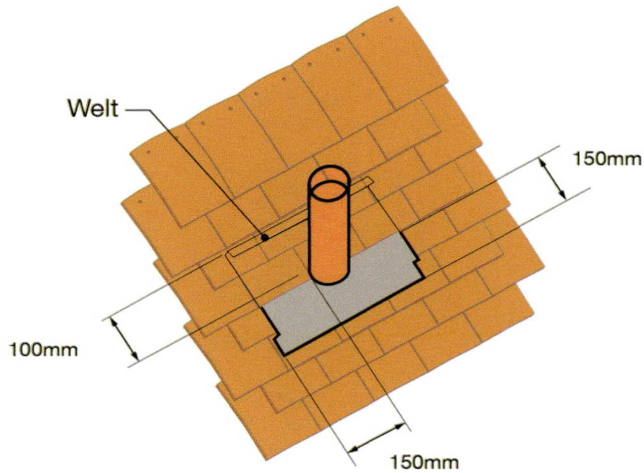

Figure 6.21 Pipe flashing

Roof window flashings

Roof windows can be supplied with the appropriate preformed flashing suitable for use with plain tiles. At the top edge, the flashing forms a top or 'back' gutter. At the bottom edge, the flashing forms a cover flashing over the course of tiles below the window. At the sides of the window, integral soakers interleave with the tile courses. (See the roof window manufacturer's details for specific installation details.)

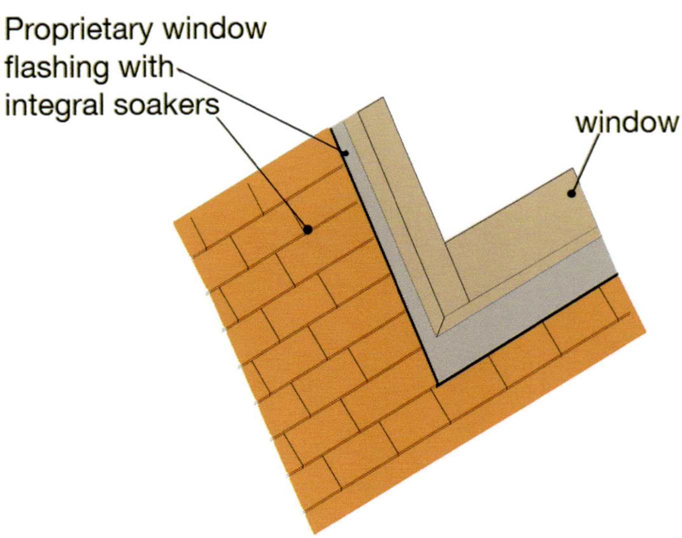

Figure 6.22 Window flashings with integral soakers

Safety hook fixings

Penetrations through the tiling such safety hooks are weatherproofed in the same way as pipes by using a suitable metal flashing. The flashing should be turned over the top of the tiles.

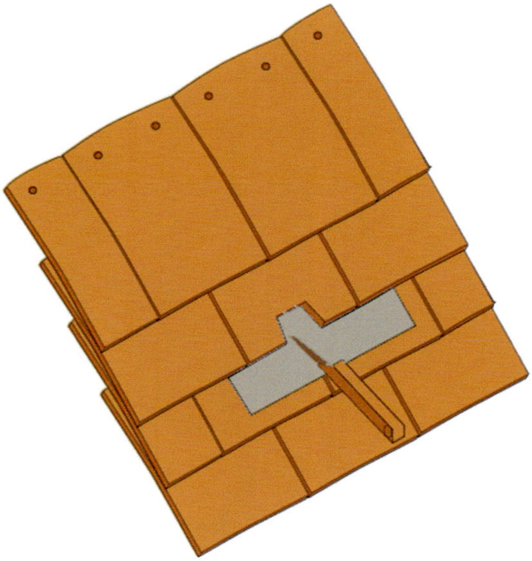

Figure 6.23 Safety hook fixings

Metal saddle flashings

Suitable metal saddle flashing should be fitted at junctions such as between ridge and hip, ridge and valley, ridge and abutments and at the top of two valleys. Figure 6.24 shows a typical saddle between the ridge and hips. The saddle should extend at least 100 mm over the top courses of tiles. A special hip stop end ridge tile should be fixed over the metal saddle (not shown for clarity).

Figure 6.24 Saddle to ridge junction

Top edge abutment

Turn the underlay not less than 50 mm up the abutment. Lay tops tiles and nail as recommended by the manufacturer. Fix tiles close to the abutment to enable a weatherproof junction to be formed by a metal apron flashing. Ensure metal flashing turns up abutment by a minimum of 75 mm. The flashing should cover the top tiles by a minimum of 100 mm. If top tiles are not used, then the flashing should be extended to 150 mm.

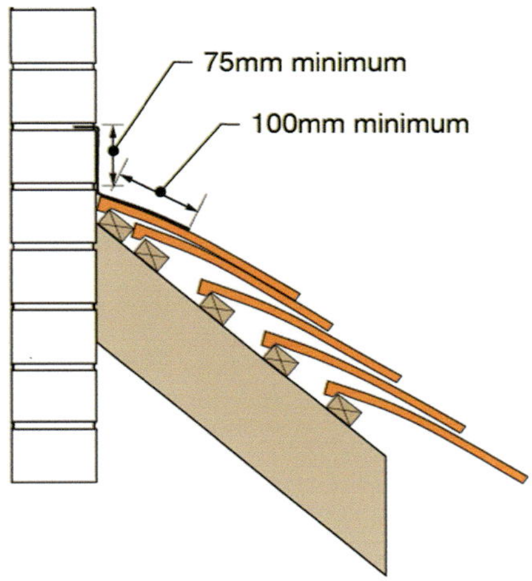

Figure 6.25 Top edge abutment

Ventilated top edge abutment

Ensure that an air gap is provided as recommended by the ventilator manufacturer. Lay top tiles and nail as recommended by the manufacturer. Fix the ventilator to enable a weatherproof junction to be formed by metal cover flashing. Ensure metal flashing turns up the abutment by a minimum of 75 mm.

Side abutment with metal soakers

Turn the underlay at least 100 mm up abutment. Cut standard tiles and tile-and-a-half tiles as necessary and interleave with metal soakers to form a close, weathertight abutment. Form soakers with a 75 mm upstand against the abutment and fix by turning down over the head of each tile. Dress a metal step flashing closely over the soakers with a lap of at least 65 mm.

6.7 Changes in pitch

There are several ways for designers to change the pitch of a roof. A mansard roof, where the lower pitch is steeper than the upper pitch in section, is a common sight. Less common is to change the pitch of the roof in plan, and a valley is usually required at the junction between the different pitches. The following chapter covers recommendations for the installation and execution requirements.

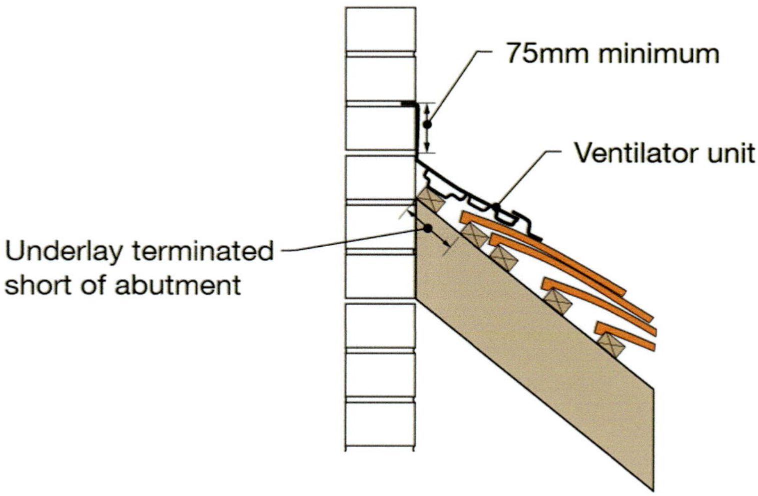

Figure 6.26 Ventilated top edge abutment

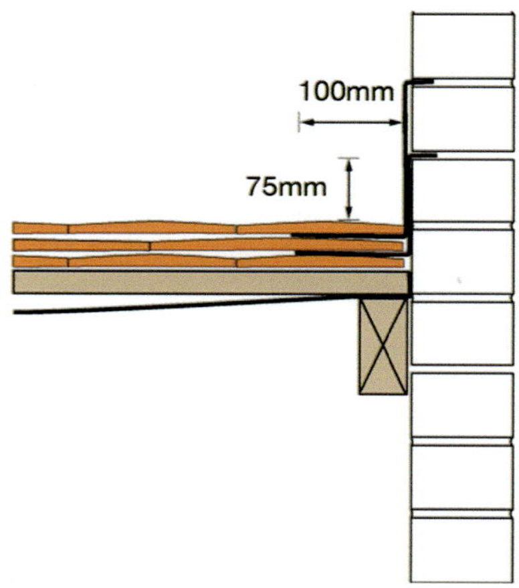

Figure 6.27 Side abutment with metal soakers

Change of pitch without flashing

This detail as shown in Figure 6.28 does not usually require a metal flashing when plain tiles are used due to their relatively small size. Generally, the tiling continues in the usual way and sweeps around the change in roof pitch. Care should be taken to maintain adequate head-lap.

Where the pitch change is significant, it may be necessary to increase the batten depth (such as using a double batten) for the first course(s) at the steeper pitch. Always ensure that the lower roof is at or above the minimum recommended roof pitch for the tiles.

Mansard with mansard tiles

Mansard tiles are purpose made to change pitch from a steep to a less-steep angle without flashings. It is important to ensure that a clear ventilation path is maintained from the eaves

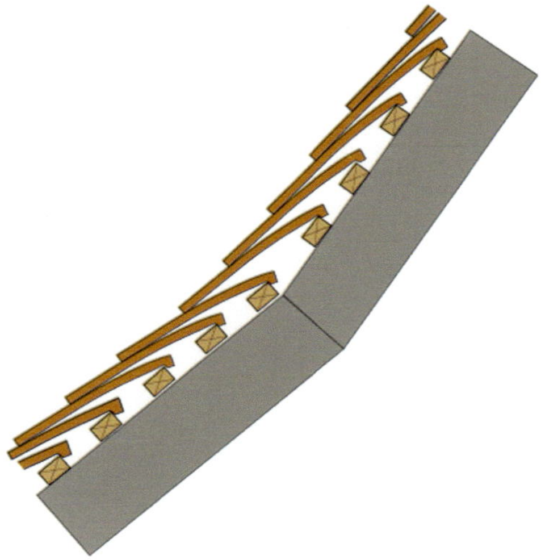

Figure 6.28 Changes in pitch

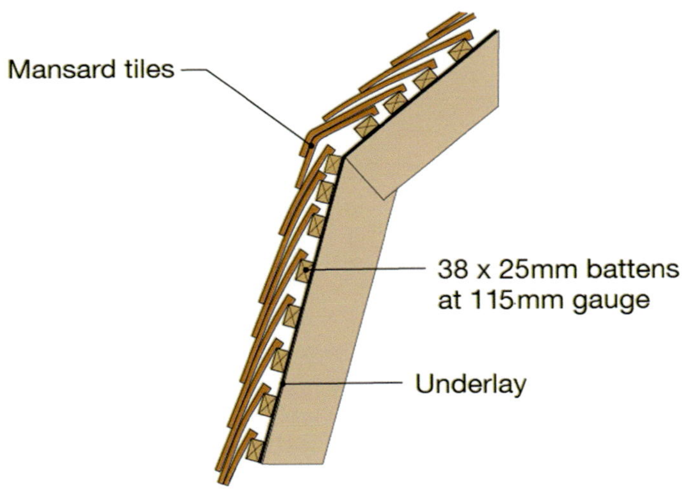

Figure 6.29 Mansard tiles

to the ridge where the insulation is positioned between the rafters. Lap the roof underlay over the mansard underlay by a minimum of 150 mm.

Set out the tiling battens to ensure that a 65 mm head-lap is maintained for the mansard tiles. Mansard tiles need to be specified for the particular roof pitches required.

Mansard with metal flashing

Fix continuous timber tilt batten at eaves to provide support for metal flashing and eaves course of tiles. Ensure that a clear ventilation path is maintained from the eaves to the ridge where insulation is positioned between the rafters. Lap the roof underlay over the metal welt and secure the mansard underlay under the tilt batten. Dress cover flashing over the top course mansard tiles by a minimum of 100 mm and extending 150 mm to 200 mm up the tilt batten. Clip the bottom edge of the flashing in exposed locations. The use of mansard tiles should be encouraged wherever possible.

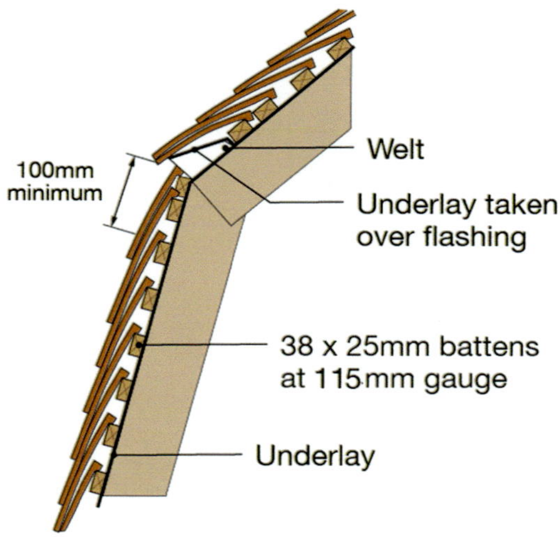

Figure 6.30 Metal flashing

6.8 Roof windows

The following chapter covers recommendations for the installation and working requirements for common roof window details. It is important to provide moisture barriers and fire stopping measures in accordance with building regulations.

Roof window side edge detail

A gap 30 mm wide should be maintained between the tiling and side edges of the window frame. All flashings and other weathering components must be fitted in accordance with the window manufacturer's recommendations. All seals and gaskets have been omitted from Figure 6.31 for clarity.

Roof window bottom edge detail

The top course of tiles below the window should be cut, if necessary, to maintain a consistent head-lap over the tiling course below. A deeper batten should be used, if necessary, to maintain the correct pitch of top course tiles. All flashings and other weathering components must be fitted in accordance with the window manufacturer's recommendations. All seals and gaskets have been omitted from Figure 6.32 for clarity.

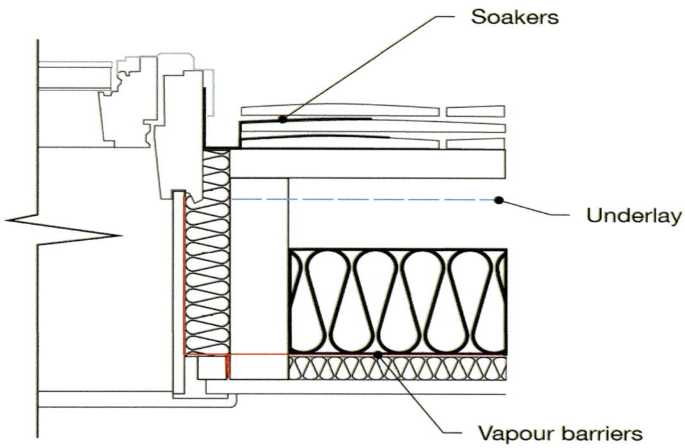

Figure 6.31 Side edge detail

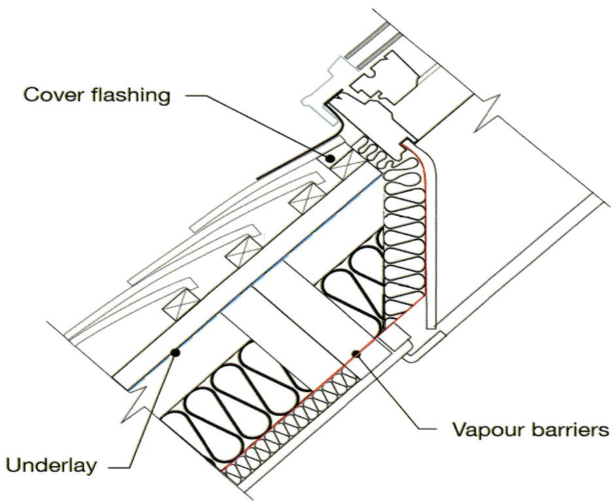

Figure 6.32 Bottom edge detail

Roof window top edge detail

A gap of 60 mm to 150 mm should be maintained between the tiling and the top edge of the window frame. If necessary, the tails of the first two courses of tiles should be cut. All flashings and other weathering components must be fitted in accordance with the window manufacturer's recommendations. All seals and gaskets have been omitted from Figure 6.33 for clarity.

6.9 Common roof constructions

Both single-lap and double-lap tiles can be used on a wide range of roof substructures. It would not be possible to illustrate every combination of roof structure and tile fixing

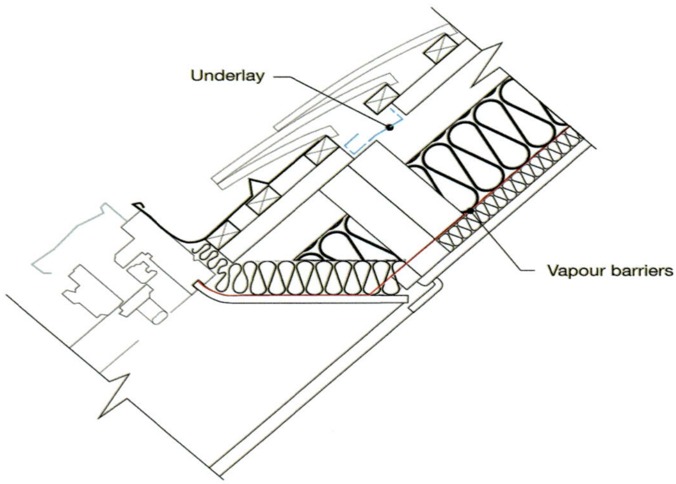

Figure 6.33 Top edge detail

methods. Nevertheless, the examples given next provide a good representative range of common pitched roof constructions.

The design of the roof substructure, such as rafters, insulated panels and so on, and the securing of the substructure to the building are not within the scope of this chapter.

Domestic building with trussed rafters, masonry walls and ceiling insulation

Insulation is laid between and across ceiling joists. Rafter roll enables a ventilation air path from the eaves into cold loft space. For an efficient well sealed ceiling, install a separate Air & Vapour Control Layer (AVCL) between the ceiling and insulation, ensuring there is a good seal at all perimeters, junctions and penetrations. Seal all gaps between the ceiling and masonry wall and seal all penetrations through the ceiling using flexible sealant.

Domestic building with attic roof, timber frame, brick outer skin and insulation between and below rafters

Tightly butt, tape and seal all joints between and around insulation boards using tape and jointing material to create an effective vapour control layer. Alternatively, install a separate vapour control layer between the ceiling and insulation, taking care to ensure a good seal at perimeters. Ensure continuity from ceiling AVCL to wall AVCL. Seal all penetrations through the ceiling using flexible sealant. Use jointing tape at the ceiling to the wall joint prior to the finish plaster being applied. Fit an air impermeable cavity closure at the top of the external wall.

Domestic building with attic conversion, timber frame, brick outer skin and insulation between and below rafters

The following diagram shows the same domestic building set-up as the diagram example shown previously (Figure 6.34), but now with an attic conversion detailed.

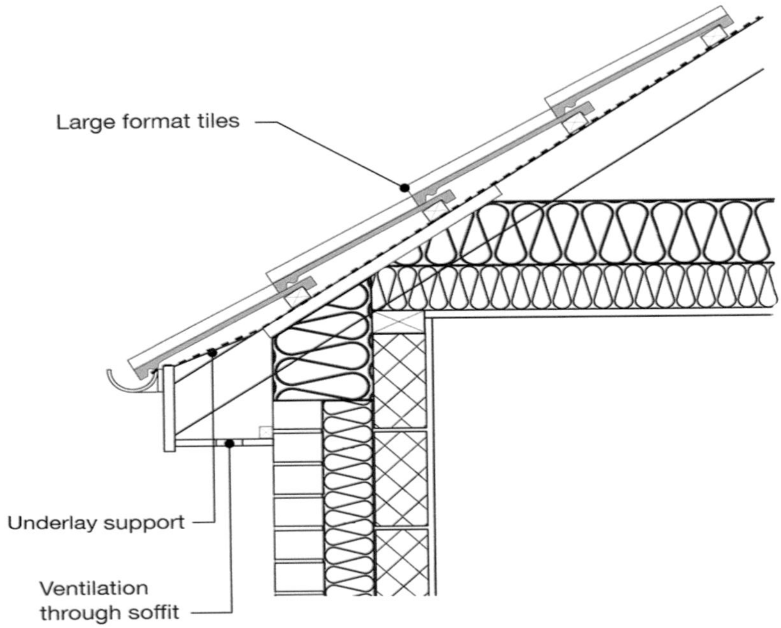

Figure 6.34 Domestic building masonry walls

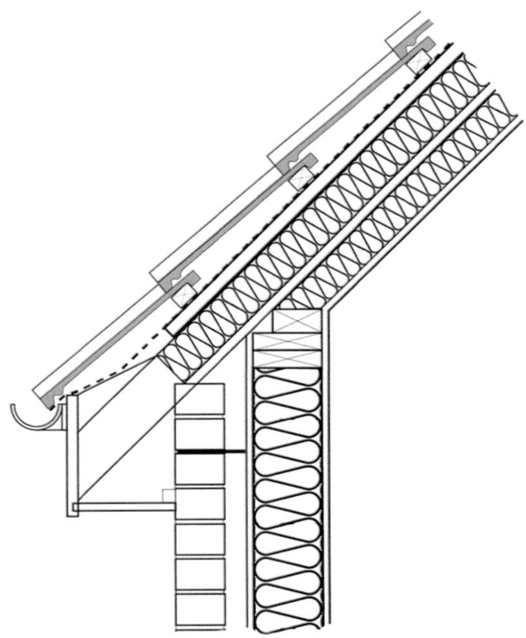

Figure 6.35 Warm roof detail

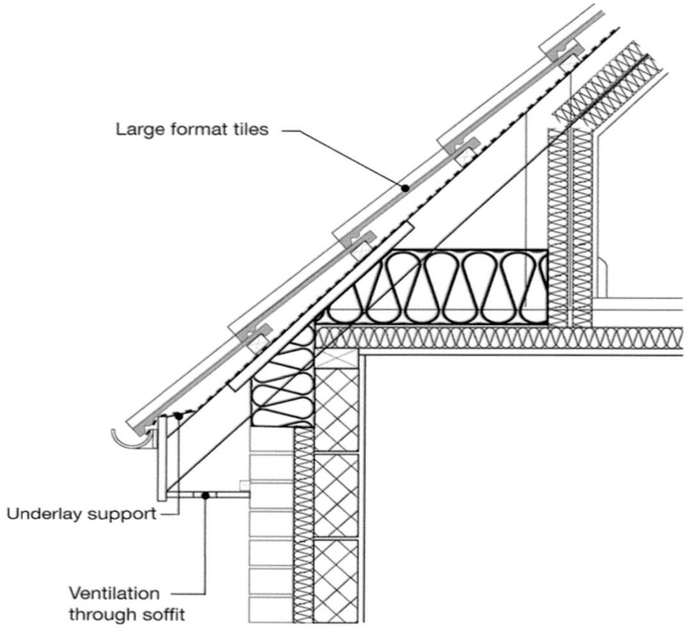

Figure 6.36 Habitable warm roof detail

Domestic building with SIPs roof panels, timber frame, brick outer skin with vertical tiling

Structural insulated panels (SIPs) with extra insulation are installed under the roof panels. Carefully seal all joints between each panel and other elements such as walls, chimney stacks and so on. Install an AVCL between the SIPs panels and plasterboard ceiling, taking care to ensure a good seal at all perimeters, junctions and penetrations. Ensure continuity from ceiling AVCL to wall AVCL.

Laps should be not less than 50 mm and sealed with adhesive or double-sided tape, ideally formed at solid supports such as joists or rafters. Seal all penetrations through the ceiling using a flexible sealant. Use jointing tape at the ceiling-to-wall joint prior to the finish plaster being applied.

Non-domestic building with SIPs roof panels, SIPs external wall frame and vertical tiling

Leave a 5 mm gap between panels and carefully seal by first injecting polyurethane foam and then applying a bead of bitumen sealant. Carefully seal all joints between the panels and other elements such as walls, chimney stacks and such like.

Install an AVCL between the SIPs panels and plasterboard ceiling, taking care to ensure a good seal at perimeters, junctions and penetrations. Ensure continuity from ceiling AVCL to wall AVCL, ideally formed at solid supports such as joists or rafters. Seal all penetrations through the ceiling using a flexible sealant. Use jointing tape at the ceiling-to-wall joint prior to the finish plaster being applied.

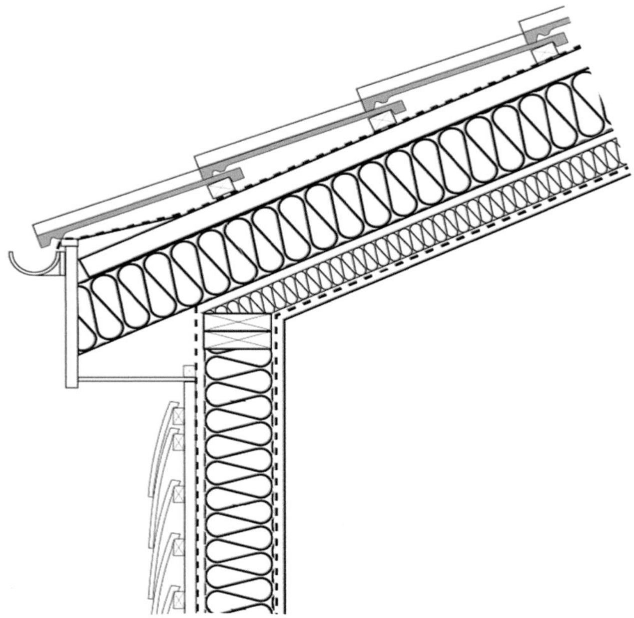

Figure 6.37 Domestic building with SIPs panels

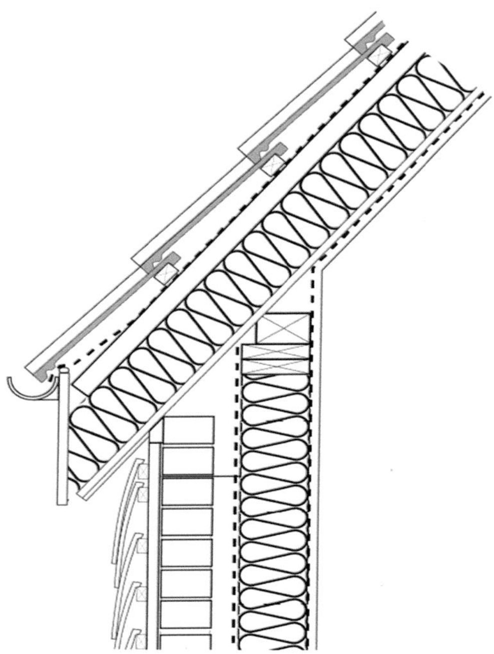

Figure 6.38 Non-domestic building with SIPs panels

Metal frame building with timber rafters, insulated Tactrays, steel frame wall and brick outer leaf

Tightly butt, tape and seal all joints using appropriate tape and jointing material to create an effective vapour barrier. Alternatively, install a separate AVCL between the ceiling and insulation, taking care to ensure a good seal at perimeters, junctions and penetrations. Ensure continuity from ceiling AVCL to wall AVCL.

Laps should be not less than 50 mm and sealed with adhesive or double-sided tape, ideally formed at solid supports such as joists or rafters. Seal all penetrations through the ceiling using a flexible sealant. Use jointing tape at the ceiling-to-wall joint prior to the finish plaster being applied.

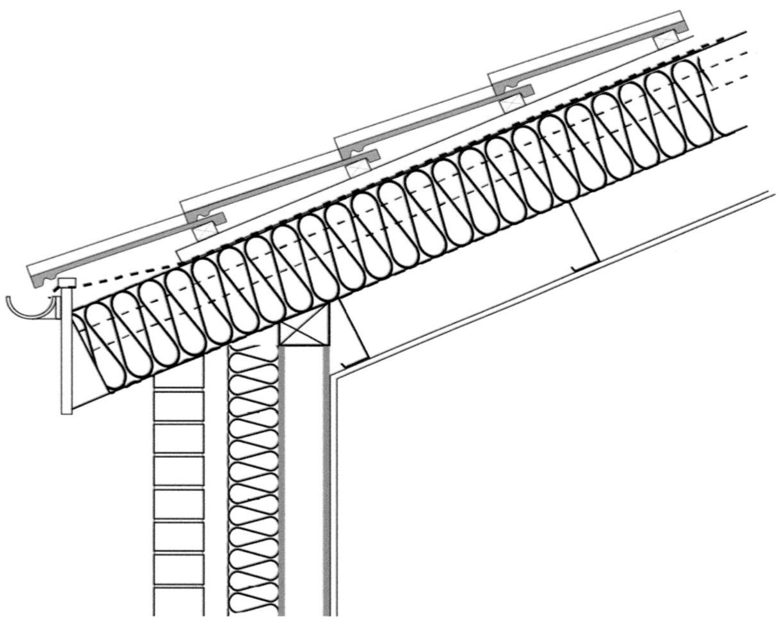

Figure 6.39 Metal-framed building

6.10 Laying large format single-lap tiles

The following information is for general guidance only and should be read in conjunction with previous chapters in this section.

The designer and fixer should ensure that tiles are installed in accordance with the Code of Practice BS 5534. Local conditions and current good practice should also be considered. All work on site should be carried out in accordance with the Code of Practice BS 8000–6.

Storage

Large format single-lap tiles are generally supplied shrink wrapped on wooden pallets and should be stored on firm, level ground.

Ventilation and dry systems

Guidance on the installation of ventilation and dry-fix systems can be obtained from the tile manufacturer. Ventilation should be installed to comply with the recommendations given in BS 5250, and dry systems should be installed to resist the maximum design wind load for the site.

Setting out the roof (general)

Battens should be secured by nailing into each rafter using galvanised or sherardised smooth round nails. Annular ring shank or helical threaded shank nails may also be used. The minimum nail size is 65 mm x 3.35 mm, but the resistance of the fixing to wind uplift resistance should be determined using the information given in BS 5534.

Each batten should be not less than 1,200 mm long. For trussed rafter roofs where the batten gauge is more than 200 mm, there should be no more than one joint in any four consecutive battens on the same rafter. Where the batten gauge is less than 200 mm, there should be no more than three joints in any 12 consecutive battens on the same rafter. Where battens are jointed, ensure the cut ends are square and the joint is located centrally over the rafter. Secure each batten end by splay nailing. Ensure that full support is provided to fix the ends of the battens at hips and valleys.

Setting out up the roof (gauge)

Set the first batten at the eaves to allow the tails of the eaves course tiles to overhang the fascia just short of the centre of the gutter. Generally, battens should be parallel to the eaves and ridge lines. Where this is not possible, due to the shape or design of the building, fix the battens to be at right angles to the line of drainage.

Set the last batten at the ridge so that the ridge tiles will overlap the top course of tiles by at least 75 mm. See Figure 6.40 drawing below.

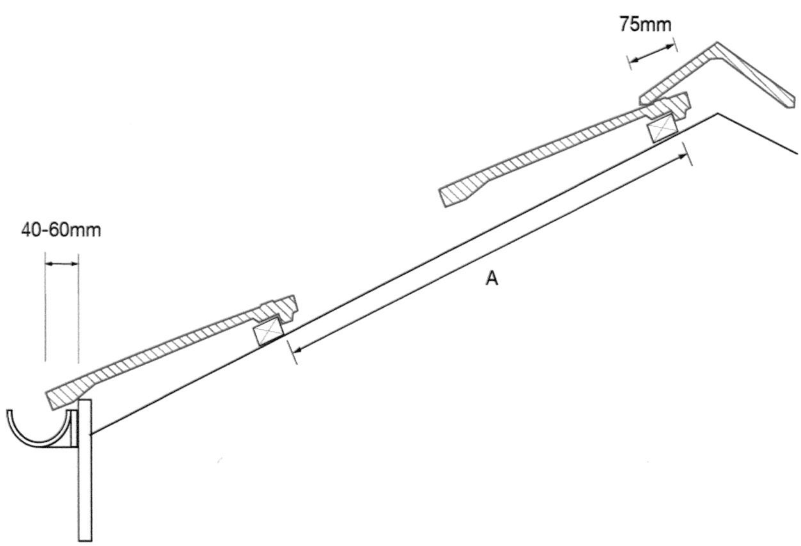

Figure 6.40 Setting out the gauge for single-lap tiles

To determine the gauge for variable lap tiles, measure the distance (A) from the first eaves course batten to the top course batten. Divide this distance by the maximum tile gauge. Round the answer up to the nearest whole number (B) and this gives the number of tile courses required.

Divide A by B to determine the actual batten gauge. Fix the remaining battens at this gauge.

Alternatively, it is technically possible to set out most of the roof at the maximum gauge and then reduce the gauge of the last few courses to avoid having a cut course at the ridge. Care must be taken if this method is used to ensure that doing this will be aesthetically acceptable.

For fixed-gauge tiles, check the batch of tiles to determine the shunt possible between courses. If it is not possible to set the battens to obtain full courses between the limits of the available shunt, then it will be necessary to have a short course of cut tiles at either the bottom or top edge.

Setting out across the roof

Lay a course of tiles along the eaves, setting the tiles at the average linear coverage. At this stage, some adjustment in shunt could be made, if necessary, to allow a 38 mm to 50 mm overhang at verges. Ensure the verge overhang is equal at the left and right ends. Mark the position along the eaves and top battens of every third tile.

Alternatively, take a gauge rod (a short length of tile batten) and mark the position of three tiles with their sidelocks fully closed, then mark the position of the three tiles with their sidelocks fully open. Set the average coverage by making a third mark midway between the previous two marks on the rod. Use this third position to mark out along the eaves and top battens.

Strike a chalk line from eaves to the ridge at each mark. Tiles can be laid to these marks to ensure perpendicular lines remain straight.

Tile fixing generally

Load out all sides of the roof uniformly, randomly mixing tiles from a minimum of three different pallets. Lay tiles starting at the right-hand side of the roof plane and working toward the left.

Use fittings supplied by the manufacturer. Make sure that every third tile is positioned to the chalk line. Mechanically fix all tiles in accordance with the manufacturer's fixing specification.

Pitch transitions

A mansard roof has two slopes on one or more of its sides, with the lower slope at a steeper angle than the upper. A bonnet roof is essentially a mansard roof in reverse. Also known as kicked eaves, a bonnet roof sides with an upper slope steeper than the lower.

When using large format tiles, where the pitch transitions, the joint is normally weatherproofed using a lead flashing detail. Examples of this are shown in Figures 6.41 and 6.42 below. Some tile manufacturers will produce purpose-made mansard tiles which can be used at the junction instead of the flashing.

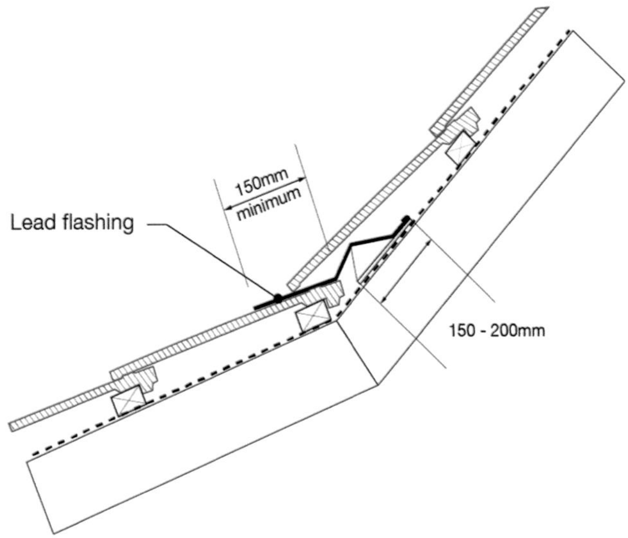

Figure 6.41 Change of pitch

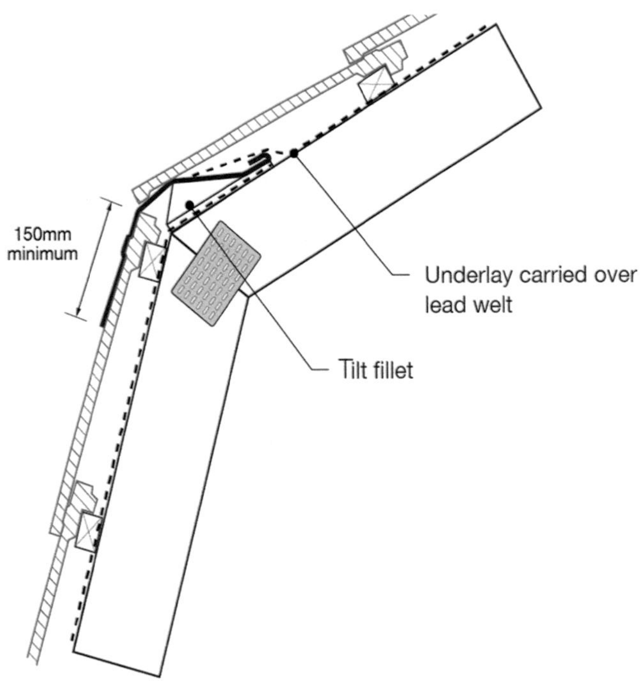

Figure 6.42 Mansard

6.11 Laying of double-lap plain tiles

This chapter is concerned with the laying and fixing of plain tiles and should be read in conjunction with previous chapters in this section, and Chapter 3.7 *Peg tiles and plain tiles*.

Plain tiles made from clay have been used to cover roofs in Britain for over eight hundred years and they form much of the character of the roofs seen in the South East of England and the Midlands, where the largest deposits of clay are located.

Nowadays, plain tiles are made of clay or concrete and come in a variety of colours. Concrete tiles are heavier, therefore, if they are used as replacements for clay, it must be determined that the roof structure is able to support the additional weight.

Storage

Double-lap tiles including plain tiles are generally supplied shrink wrapped on timber pallets or in timber-slatted crates. They should be stored on firm, level ground no more than two pallets high, but preferably single height.

Points to note

- **Pitch** – Clay plain tiles conforming to the dimensional tolerances given in EN 1304 can be laid on rafter pitches down to a minimum of 35°. Plain tiles which, for aesthetic reasons, do not comply with the dimensional tolerances given in EN 1304 must be laid at pitches not less than 40°. Roofs with plain tiles generally have a higher pitch than large format tiles.
- **Head-lap and side-lap** – The head-lap specification for plain tiles is 65 mm minimum and the maximum gauge of battens should be 100 mm. Gauges of less than 88 mm are not recommended. The side-lap should be not less than one-third the width of the tile, typically 55 mm.
- **Securing the tiles** – The minimum fixing specification for plain clay tiles is to fix every fifth row, but in all cases, uplift calculations should always be completed to ensure that the specification meets the wind load requirements.
- **Weather resistance** – The double-lap method of laying tiles as described in BS 5534 provides an excellent rain and snow protection system.
- **Fire resistance** – Clay tiles satisfy the building regulations with respect to external fire performance because they are incombustible and have a reaction to fire rating better than Class A2 within BS EN 13501–1.
- **Ventilation** – Due to the camber (curvature) of plain tiles, they do allow an amount of natural ventilation. Ventilation can be provided by the use of purpose-made vent tiles, or third-party products fitted in place of tiles. They should be installed to comply with the recommendations given in BS 5250.

Setting out up the roof (gauge)

The setting out for double-lap plain tiles is done using the same base principles as those used for the setting out of single-lap large format tiles. The main difference is that due to the smaller size of the tiles, and the double-lap laying design, many more battens are needed.

The main points to note are:

- Determine the position of the battens for the bottom row of eave tiles, and the bottom row of full tiles, allowing for the overhang of the tiles into the gutter.
- Determine the position of the uppermost batten for the top row of tiles. Typically the top of the uppermost row needs to be at least 65 mm under the edge of the ridge tile.
- Take the measurement between the top of the upper batten and the top of the lower batten and divide this by the manufacturer's maximum gauge requirement to give the number of tiles required. Round it up to the next whole number.
- Now divide the distance between battens by the rounded number and you have the gauge actually needed.

If you have a window or a dormer in the roof, these present more 'fixed points' as well as the bottom and top of the roof. For example:

- under the sill of a window
- at the top of the window

Therefore, the gauge needs to be determined with these additional fixed points:

- From the bottom row eaves batten to the batten under the window sill.
- From the batten under the window to the top of the window.
- From the top of the window to the top row batten by the ridge.

Knowing these possibly different gauges, it can be determined if the gauge needs adjusting. If it does, the decision can be made as to how to do this and keep the minimum head-lap required, plus achieve the best aesthetics for the roof. If different gauges are used between the fixed points, these need to be used also on adjoining roof slopes to keep the tile courses in line.

Setting out across the roof

Having set out and nailed all the battens the setting out of tiles on each batten needs to be considered to ensure perpendicular lines remain straight. Each plain tile is 165 mm wide and should have a gap of up to 3 mm between its sides. This gap is often called the 'shunt gap':

- Lay a course of tiles along the eaves starting with between 38 mm and 50 mm overhang at the verges.
- Equalise the spacing of the tiles for the 'shunt' gap between adjacent tiles.
- Mark the position of every third tile along the batten, and repeat along the top batten.
- A perpendicular line can be dropped from the top to the bottom eaves battens and the battens going up the roof can all be marked.

An alternative method is to use a gauge rod. This is a short length of batten marked with the average coverage of three tiles when laid:

- The gauge rod is marked at the midpoint between three tiles with side gaps fully closed and three tiles with side gaps fully open.
- Use this midpoint position to mark out along the eaves and top battens, remembering to allow for the overhang at the verges.
- A perpendicular line can then be dropped from the top to the bottom markings and the straight lines can be marked out.

Tile fixing generally

Load out all sides of the roof uniformly, randomly mixing tiles from at least three different pallets.

Lay tiles starting at the right-hand side of the roof plane and working toward the left. All verges are to be laid broken bond with full tile and tile-and-a-half in alternate courses. Depending on fixing specifications, the third and fourth tiles from the left-hand verge may be left out to enable the use of tile battens as a ladder to the ridge.

Make sure that every third tile is positioned to the chalk line, and mechanically secure all tiles in accordance with BS 5534 fixing specification.

The eaves course of tiles is laid broken bond to the first full course of tiles so you may need an eaves tile-and-a-half. These can be made by cutting a full tile-and-a-half down to the same length as the eaves tile. The same may apply at the top tile course and is more critical since it is more visible.

At roof window and dormers reveals and valleys, it may be necessary to finish with a cut tile-and-a-half on each course to maintain the vertical perpendicular joints.

Manufacturers always recommend following the guidelines contained in the Code of Practice for Slating and Tiling BS 5534, and Code of Practice for Workmanship BS 8000–6.

CHAPTER 7

Specialist tiling techniques

7.1 Ornamental tiles

The following chapter includes drawings that illustrate some of the range of decorative finishes that are available. The drawings only deal with the physical shape of the tiles and do not cover the wide range of colours that are available.

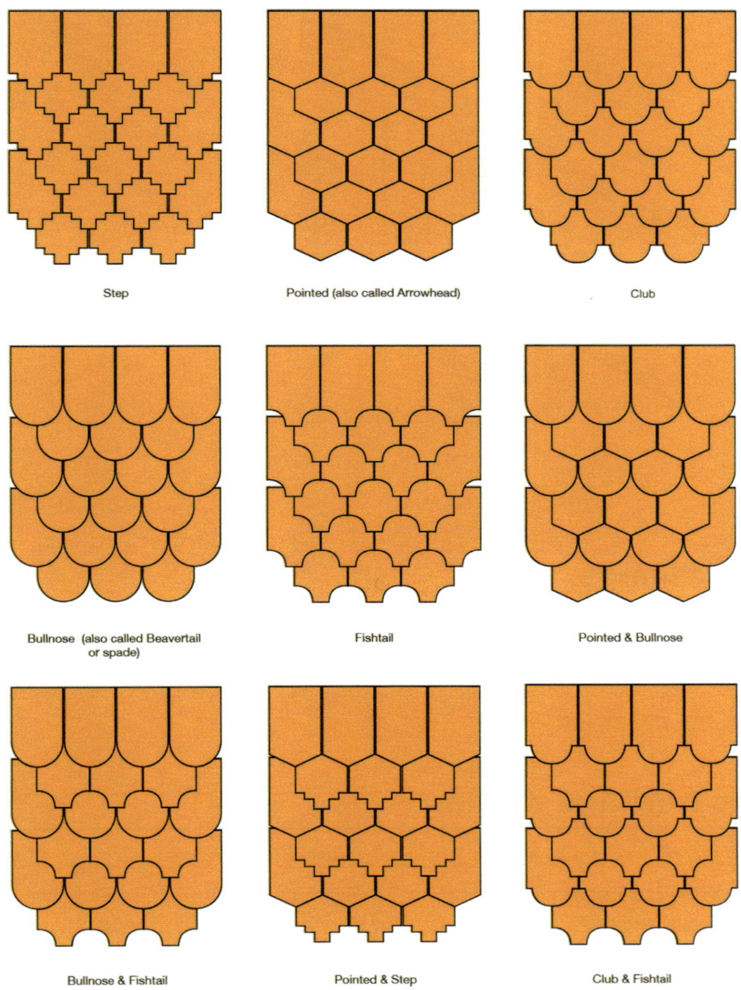

Figure 7.1 Ornamental tiles examples

DOI: 10.1201/9781003196990-7

Mathematical tiling

Mathematical tiles are used for the external cladding of walls and give a finished appearance of bricks. Nowadays, these are only used on historical buildings and are a specialist bespoke-made product.

Fix 38 mm x 25 mm tiling battens to the specified gauge for the design of the product and secure to groundwork with recommended fixings. Lay tiles in broken bond fashion and twice nail each tile. Joints may be left dry or can be bedded and pointed in accordance with the manufacturer's instructions.

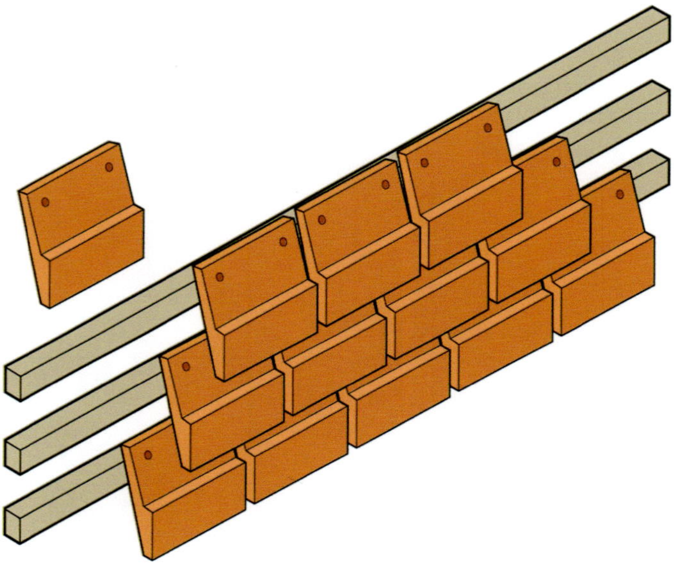

Figure 7.2 Mathematical tiling

Decorative flashings

Where a top course tile is used, the minimum coverage of the flashing is 100 mm. If no top tile is used, coverage should be extended to 150 mm. The decorative section of flashing should be added onto the coverage such that it does not compromise the lap relative to the vertical joint and the nail holes.

7.2 Historic heritage roofing

The UK is fortunate to have a wealth of historic, well-preserved buildings that attract visitors from around the world. In order to ensure that these properties continue for generations to come, a considerable amount of work and finance has to be invested.

In terms of roofing, this means that occasionally the original roof tiles need to be replaced with good-quality reclaimed tiles or new ones to protect and maintain the appearance and

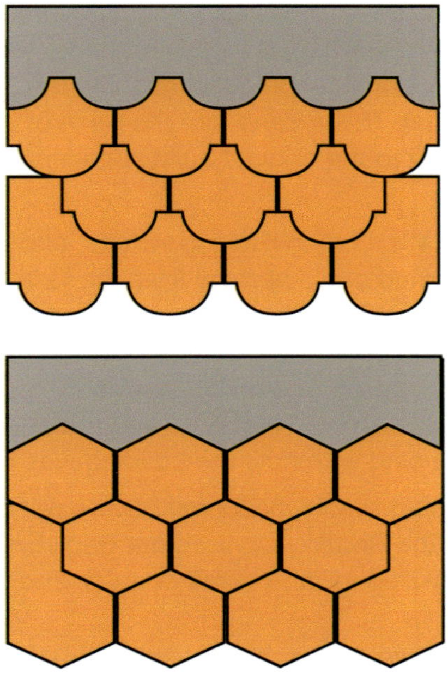

Figure 7.3 Decorative flashings

fabric of these valuable properties. However, these heritage projects are often far more complex than installing standard roofs on modern housing projects. Manufacturers, clients, architects, UK heritage agencies and planning departments all work closely together to achieve the desired finished result.

Using the correct materials is vital to all heritage projects, as strict regulations often require roof replacements or repairs to be carried out with like-for-like products. New roofing tiles with the correct profile, colour and texture are needed to match existing tiles, the age and style of a building and the surrounding environment.

The heritage roofing sector is dominated by clay products, and it is common for bespoke tiles to be specially made for projects where the original roof design needs to be strictly adhered to and mass-manufactured products are not suitable. After clay, it is also typical for natural stone, natural slate and thatch to be used on heritage projects.

The challenge does not end with sourcing the right roofing tile. An immense amount of planning and preparation goes on behind the scenes with heritage projects as attention to detail is paramount. Correct installation is also as important as choosing the correct product to work with.

Roofing contractors who have experience with historic buildings exhibit a high level of craftsmanship as heritage roofing is very methodical and has to be completed to strict high standards. Conservation boundaries also need to be considered, with appropriate analysis and documentation completed.

Modern roofing tiles have been developed with design features that make installation as quick and simple as possible. Historical (usually handmade) roof tiles are often more difficult to fit and therefore require traditional techniques, such as the bedding and pointing of hips, ridges and verges in lime mortar. This means that the pace of work is much slower, but an exceptional finish that is faithful to the original build is considered to be more important than a quick installation.

Some heritage roofing projects will be on listed buildings, and the level of the listing will dictate what works can be carried out. Approval from the local planning department and/ or heritage agency should be sought before starting any works. These regulations act as a work method guide to heritage and restoration projects to ensure that the work is undertaken and finished to an appropriate result and standard. For example, plastic clips should not be used, as these are an incorrect type and material for these buildings. The method used should be in keeping with how the building was built originally.

When seeking roofers for these projects, historical bodies will often look for someone who is listed on the National Heritage Roofing Contractors Register, which is managed by NFRC. These are some of the most skilled roofers in the country, and crucially learnt their skills on historic buildings.

Taking inspiration from the UK's historic buildings, there is a growing trend for new build constructions adopting the look of heritage projects by using products such as handmade roof tiles. Customers may often prefer the look of a roof that appears to be aged and weathered rather than a roof that is very obviously brand new.

Laying of historic peg tiles

This chapter is concerned with the laying and fixing of peg tiles and should be read in conjunction with Chapter 6.3 *Setting out battens and tiles*. For information on the types of peg tiles, holes, fittings and substructure, please see Chapter 3.7 *Peg tiles and plain tiles*.

Restoration specification comprehensive

Where true peg tiles are to be used or reused, it is necessary to decide whether the appearance of the underside of the roof is important. The external appearance of a tiled roof will not normally betray whether the tiles are pegged or nibbed. The specification will therefore be based on the efficiency of the finished roof and the economical use of the materials available. Where it is so deemed necessary to use a traditional specification, the following example would be appropriate.

Example

Peg tiles are historically laid to 60 mm head-lap with secure (wooden) pegs, one per tile. These hook over split (hardwood) laths that are nailed to rafters with a 3.35 mm gauge, round-head, galvanised, steel-wire nails.

Supplementary fixing may be obtained by specifying that the tiles should be torched with lime sand hair mortar. Sand lime hair mortar torching, when correctly applied, will also act as bedding at the head of the tile. Where specified, a suitable vapour permeable membrane may be used in place of the torching.

An alternative to torching as a supplementary fixing is that peg tiles should be nailed to rafters where peg holes coincide, using 3.35 mm diameter clout head copper wire nails.

Where peg tiles have very large holes, it may be necessary to specify 'extra-large clout head' 3.35 mm diameter wire nails for this purpose. In some districts, normal clout head nails are used with 15 mm to 20 mm washers for extra-large peg holes.

These methods of nailing are preferable to machine drilling or grooving of clay tiles where the shock to the case hardening is disadvantageous and should be avoided, particularly where the original tiles have already been exposed to years of weathering.

Restoration specification (external appearance only)

Example

Existing peg tiles are to be laid to 65 mm head-lap, and hooked with 4.5 mm (or 5.0 mm) diameter x 30 mm (or 40 mm) long aluminium round point tile pegs, using one per tile. The battens should be 38 mm x 19 mm (or 25 mm) treated sawn softwood and an appropriate underlay used.

Supplementary fixing may be obtained as stated earlier (not torching) or as follows. Each tile in every fourth (or fifth) course shall be twice skew nailed through the peg hole into the top of the tile batten, using 30 mm or 40 mm long x 2.65 mm diameter clout head copper wire nails.

The nail head, when skew driven, should occupy the peg hole without pulling right through. In case of difficulty, the diameter of the nail should be adjusted, remembering that a clout head is three times the diameter of the shank (see BS 1202).

Alternative supplementary fixing may be achieved by spot bedding each tile in every fourth course to the head of the tiles in the course below in sand and lime cement mortar. If it is desirable to increase the number of courses of tiles bedded, the maximum amount of bedding of tiles should leave alternate courses without bedding.

Cement mortar bedding should be at the top of the tile and kept within the head-lap. The side-laps should be kept clear (see BS 8000–6).

Bedding and torching

Where tiles are bedded for security, a suitable cement mortar should be used to avoid frost erosion. Lime mortar torching should contain hair for similar reasons.

General considerations

In most circumstances peg tiles should not be tight nailed to any type of lath or batten due to the amount of 'cocking up' of the bottom edge of the tile due to the unevenness of the tile, and which will not subsequently settle down. Tiles of more even manufacture, which have smaller nail holes rather than larger peg holes, may be either nailed to battens or better still, hung using non-rusting metal pegs.

Where a combination of pegging several courses and nailing an intermediate course is adopted, it is often possible to nail the lower of the two holes (where holes are offset) and hang using the higher one. Where double nailing is required or nail holes are not offset, the batten to each nailed course will have to be fixed higher in order to maintain even laps and margins. The amount of nailing higher would normally be 15 mm to 20 mm.

Battening to oast houses is often performed by double layering thinly cut battens vertically in order to follow the curve of the roof. Timber relatively free from knots will be needed. At the tightly curved top of the rafters, the double layer of battens will stand vertically away from the rafters. In lower positions, where the curvature is not so great, the double layer of battens will be nailed more closely to the rafter. The tensions in the battens and the lack of substance for nailing the tiles have led to the general recommendation that jagged or ring-shanked nails should be used.

Oast houses are tiled with a mixture of normal-sized tiles combined with the three sizes of tiles tapered on both sides. The tapered tiles will normally have one nail hole at the top edge centrally.

Care should be exercised in reusing peg tiles so that tiles from different areas, or of varying sizes or ages, are not mixed together. Where supplementary modern tiles are to be used, a careful comparison of shape, texture and colour should be made and, if possible, the new tiles should be used on a selected roof area so that setting out and fixing may be adjusted to suit. Likewise, the use of tile-and-a-half (extra wide) tiles at gables or for cutting, if not very carefully matched, may result in obvious disfigurement.

7.3 Vertical tiling guide

The construction of vertical tiling has many similarities with pitched roof tiling, but there are some important differences that are explored in this chapter.

Historically, the use of plain or peg tiles for vertical tile hanging is well established within the UK as it was an effective way of cladding a timber frame-built property to achieve a robust, weather-resistant and long-lasting external finish. In more recent times, vertical tiling may be carried out in a contrasting colour to the roof to obtain the unique visual impact that this achieves.

The means of securing the counter battens or battens to the wall structure should be decided before the tile battens are set out and nailed to the counter battens or the wall structure. It is important to provide moisture barriers and fire-stopping measures in accordance with building regulations.

Fire safety

Following the tragic events of the Grenfell Tower fire in 2017, there has since been a tightening of building safety laws regarding cladding external walls. Amendments have been made to both the building regulations and to Approved Document B, and a new Building Safety Act has been introduced.

For relevant buildings over 11 metres in height, the materials which become part of an external wall or specified attachment achieve not less than European Class A2-s1, d0 classified in accordance with BS EN 13501–1. There are some exceptions for membranes, and both concrete and clay tiles are fire-safe, but wooden battens do potentially present a problem. The use of fire barriers and fire stops is required as advised in BS 8000–6.

Types of wall construction

It is possible to hang tiles on almost any wall construction, but it would not be possible to illustrate every combination of wall structure and tile fixing method. The most common construction materials for walls to be cladded with tiles are as follows:

- Lightweight concrete blocks.
- Dense concrete blocks.
- Stone.
- Old and new common bricks.
- Precast concrete panels.
- Timber stud and plywood sheathing.
- Metal frame.
- Rendered walls.

Batten and counter batten fixing

BS 5534 and BS 8000–6 recommend that both an underlay and counter battens should always be used for vertical tiling. The use of underlay isolates the timber counter batten from any damp in the wall materials, and the use of counter battens, rather than just fixing tiling battens directly into the surface of the wall, minimises the number of fixings into the wall as well as creating an air gap and drainage channel behind the vertical tiles.

The methods for securing battens and/or counter battens to the wall may be influenced by the age of the wall as well as the build material. Older walls usually require counter battens secured with a proprietary corrosion-resistant fixing system, such as wall plugs and screws or shot-fired bolts.

Counter battens should have a minimum depth of 38 mm so as to accept the minimum batten nail depth of 65 mm for galvanised smooth batten nails but in all cases, the depth will be subject to wind loading considerations and the length and type of batten fixings required in accordance with calculations to BS 5534.

Counter battens are recommended for wall constructed from:

- dense concrete blocks
- precast concrete panels
- old bricks
- old lightweight concrete blocks
- stone

It is possible to secure tile-hanging battens directly to the following wall types, but best practice is to use counter battens:

- new common bricks
- new lightweight concrete blocks
- timber stud and plywood sheathing
- metal frames

Advice on fixing requirements should be sought from the fixings manufacturer and will be dependent on the material used to construct the wall.

The means of securing the counter battens or battens to the wall structure should be decided before the tile battens are set out and nailed to the counter battens or the wall structure. For masonry walls, the most common method is to install the underlay first, then fix counter battens over the underlay secured into the wall using plugs and screws or other suitable fixings and then to install the battens as normal. For timber-framed vertical areas, the battens are normally fixed directly to the timber frame.

General wall considerations

Walls should be covered with a suitable breathable (LR) underlay lapped 75 mm horizontally and 150 mm vertically. Ensure the distance of fixing at the edge is not less than 50 mm from the edge of the tiles.

Counter battens should be fixed to the wall at 450 mm to 600 mm centres. Tiling battens, with a minimum length of 1200 mm, should be fixed at a maximum of 115 mm gauge (for plain tiles).

Lay plain tiles in staggered bond ensuring tails of tiles align and fix each tile to timber tile batten with two 38 mm x 2.65 mm corrosion-resistant clout-head nails or screws.

Wall construction – blocks, bricks, stone and masonry

Lay breathable (LR) underlay (as approved) over the wall, lapped 75 mm horizontally and 150 mm vertically. Ensure the distance of fixing at the edge is not less than 50mm from the edge of the tiles.

Counter battens should be spaced at a maximum of 600 mm centres using proprietary corrosion-resistant fixings. Secure to the wall with plugs or screws or other suitable fixings. Ensure that the length and type of fixing has adequate withdrawal resistance in the material to which it is being nailed/screwed.

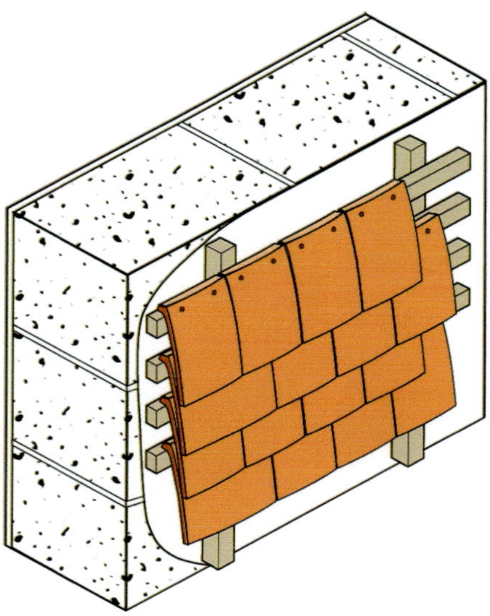

Figure 7.4 Blocks, bricks, stone and masonry

Fix tiling battens of at least 1200 mm in length to the counter battens using round wire, galvanised or improved nails of sufficient length to resist predicted withdrawal loads.

Wall construction – timber stud and plywood sheathing

Cover plywood sheathing with an approved breathable (LR) underlay. Fix counter battens over the underlay and plywood to timber studs or framing, spaced at a maximum of 600 mm centres using proprietary corrosion-resistant fixings.

Fix tiling battens to the counter battens using round wire, galvanised or improved nails of sufficient length to resist predicted withdrawal loads.

Wall construction – metal frame

Cover the sheathing panel with a breathable (LR) underlay (as approved). Secure counter battens to the metal frame using corrosion-resistant self-tapping screws or proprietary fixings as recommended by the manufacturer. Fixings should be of sufficient length to resist predicted withdrawal loads.

Setting out of battens

The setting out of the tile battens needs to consider the top and bottom of the wall, the apex (if at a gable end) and any openings through it, such as windows. Each of these points can be regarded as fixed points, this means the batten positions are in set positions and the batten gauges between them may need to be reduced to fit. Normally with plain tiles, this is done by reducing a few courses by 5 mm rather than by spreading the difference over all the courses.

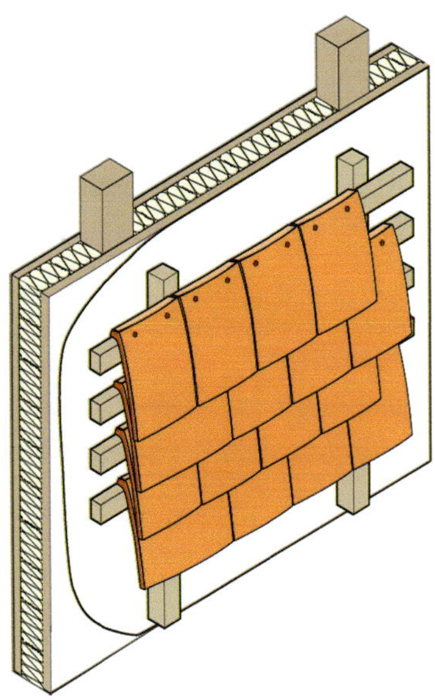

Figure 7.5 Timber construction

Figure 7.6 Metal frame construction

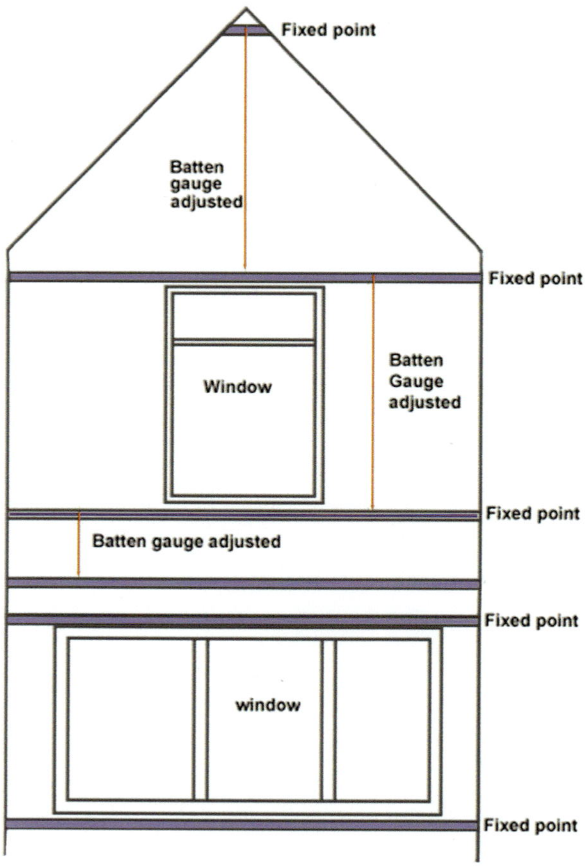

Figure 7.7 Identifying fixed points

For example, if the battens need to be dropped by a distance of 30 mm, this would normally be achieved by reducing six courses by 5 mm from the normal 115 mm gauge to 110 mm.

- Fixed point 1 – The first fixed point will be at the bottom of the vertical tiling section.
- Fixed point 2 – The second fixed point should be identified, which may be a window sill. The fixed point would be underneath the sill, and here, the batten should be set so that a full tile fits underneath the sill if being covered by a 150 mm flashing, or it may be that a full tile with an eaves/tops tile fits in better if a 100 mm flashing is used.
- Fixed point 3 – The next fixed point would be the top of the window and would be positioned so that the bottom edge of a full tile projects just over the tilt fillet/drip but does not interfere with the opening of the bay windows.
- Fixed point 4 – The next fixed point would be underneath an upper window, and here, the batten should be set so that a full tile fits underneath the window if being covered by a 150 mm flashing, or that a full tile with an eaves/tops tile fits in if being covered by a 100 mm flashing. At this point, the distance between the top edges of the battens can be measured and gauges reduced in 5 mm increments if required.
- Fixed point 5 – As with fixed point three, the positioning would allow a tile to hang just over the tilt fillet.
- Fixed point 6 – at the apex, the position of the top batten (or a board) is fixed as high as possible but in such a way that a tile-and-a-half can be nailed into place so as to finish the detail. Again, the batten gauge can be reduced in 5 mm increments if required.
- All tile battens should be horizontal (level) and straight, with no sags.

Having set out and nailed all the battens, the setting out of tiles should be considered. In general, the main goal of vertical tiling is to achieve symmetry, so centre lines are often struck and the tiling is laid from that line each way out towards the sides. For larger areas, it may be necessary to mark out every five or six tile-widths along the eaves and at high levels, and strike lines between them to keep the tiling joints aligned. The tiles can be laid touching but gaps up to 3 mm are permitted to consider any slight differences in widths.

At window reveals it may be necessary to finish with a cut tile-and-a-half on each course to maintain the vertical perpendicular joints. Some clay tiles may need to be sorted to mix oversize and undersize tiles together to ensure that an averaging out allows the perpendicular joint lines to be kept vertical. If this becomes a problem, it may be necessary to utilise the 3 mm gaps and/or trim every fifth tile down to suit.

Tiles should always be sorted and mixed from at least three pallets to ensure that the variations in shade and colour from different parts of the kiln do not give a patchy effect. Any tiles that are slightly twisted may kick out in one position but when tried in another position they sit happily.

Vertical tiling often needs more time spent during the setting out phase than does setting out pitched roofs. Being more visible, particularly at eye level, requires greater care in maintaining the vertical joint lines.

Vertical tiling to eaves with soffit

Finish battens below the soffit board to allow for a full tile and an eaves/tops tile supported on the batten to be installed. Install a lead cover flashing to lap over the top tile by a minimum of 100 mm to prevent water ingress. If no top tile is used, the lead flashing coverage should be extended to 150 mm. Nail the bottom edge of the lead flashing (at least

200 mm wide, Code 4) along the face of the top tile batten and dress it into the top tile nib space below the soffit board. Once the top tiles have been nailed, the flashing can be carefully dressed onto the surface of the top tile.

When using a decorative flashing, add the decorative section onto the coverage such that it does not compromise the lap relative to the vertical joint and the nail holes. This arrangement is useful when fixing vertical tiling to the face of an existing building.

Vertical tiling to window head: version 1

Set the first course above the window to ensure the tiles project over the window but do not interfere with the opening of the windows. Set the gauge of the vertical tiling between the window sill and window head to eliminate the need for cutting the eaves course of tiles.

Fix a shaped timber tilting fillet to the face of the lintel (approximately 100 mm x 75 mm) large enough to support the eaves course of tiles in the same plane as the rest of the vertical tiling.

Dress the underlay over the tilting fillet to a slight fall and terminate the counter battens just below the eaves tile batten to allow the underlay to dress out over the tilting fillet. Fix at least 6 mm fire-resistant board to the underside of the tilting fillet, or, as an alternative, cut plain tiles with their nibs removed could be used.

Vertical tiling to window head: version 2

Set the first course above the window to ensure the tiles project over the window but do not interfere with its opening. Set the gauge of the vertical tiling between the window sill and window head to eliminate the need for cutting the eaves course of tiles.

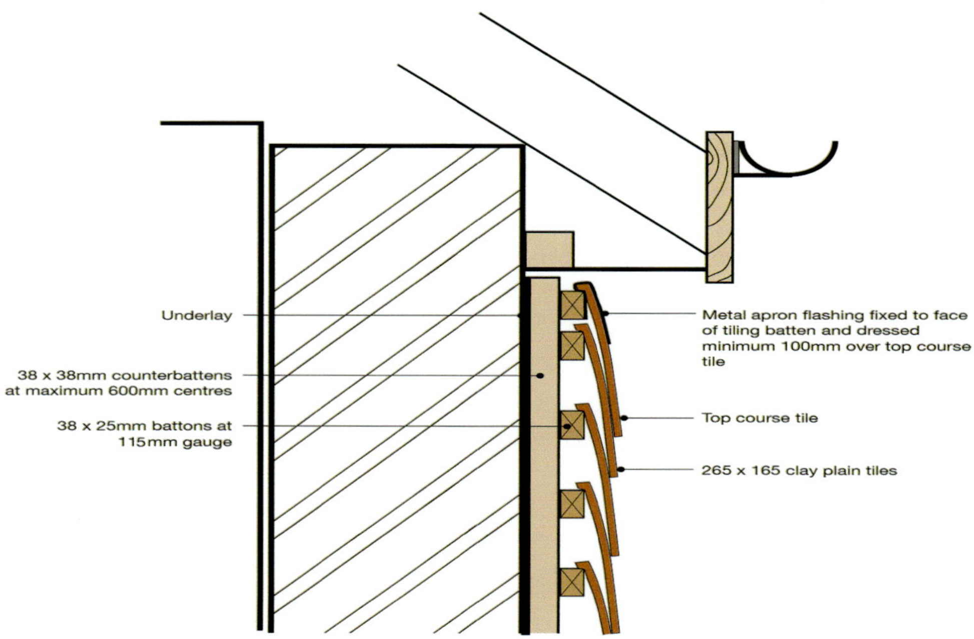

Figure 7.8 Vertical tiling to eaves with soffit detail

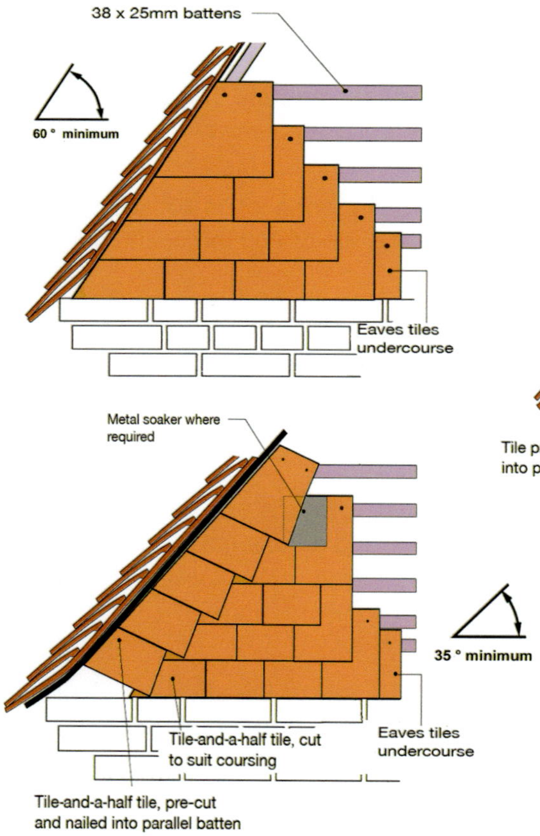

Figure 7.28 Winchester cutting for differing roof pitches

Figure 7.29 Sussex cutting

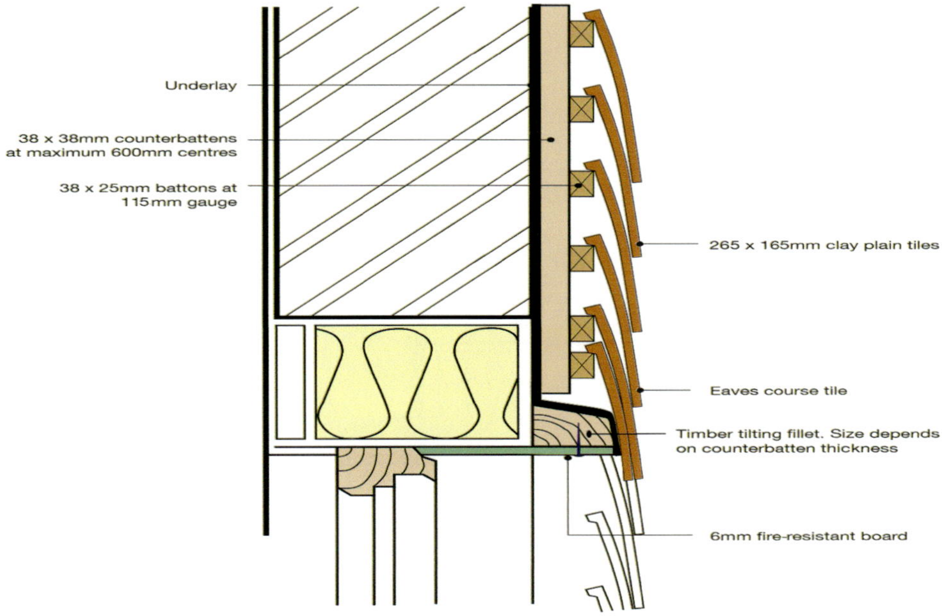

Figure 7.9 Vertical tiling to window head detail version 1

Fix a shaped timber tilting fillet to the face of the lintel (approximately 100 mm x 75 mm) large enough to support the eaves course of tiles in the same plane as the rest of the vertical tiling.

Dress the underlay over the tilting fillet to a slight fall and terminate the counter battens just below the eaves tile batten to allow the underlay to dress out over the tilting fillet. Fix a minimum of 6 mm thick fire-resistant board to the underside of the tilting fillet.

Vertical tiling to timber window sill

Finish battens below the window opening to allow the top course of tiles to fit under the windowsill. Nail the bottom edge of a lead flashing (minimum 200 mm wide: Code 4) along the face of the top tile batten and dress it into the top tile nib space. Once the tops tiles have been fixed the flashing can be carefully dressed onto the surface of the top tiles. When using a decorative flashing, add the decorative section onto the coverage such that it does not compromise the lap relative to the vertical joint and the nail holes.

The flashing should cover the top tiles by a minimum of 100 mm. If tops tiles are not used, then the flashing should be extended to 150 mm. It should extend a minimum of 150 mm beyond the jamb on either side.

Vertical tiling eaves

Fix a shaped timber tilting fillet to the face of the lintel (approximately 100 mm x 75 mm) large enough to support the eaves course of tiles in the same plane as the rest of the vertical tiling.

Dress the underlay over the tilting fillet to a slight fall. Terminate the counter battens just below the eaves tile batten to allow the underlay to dress out over the tilting fillet.

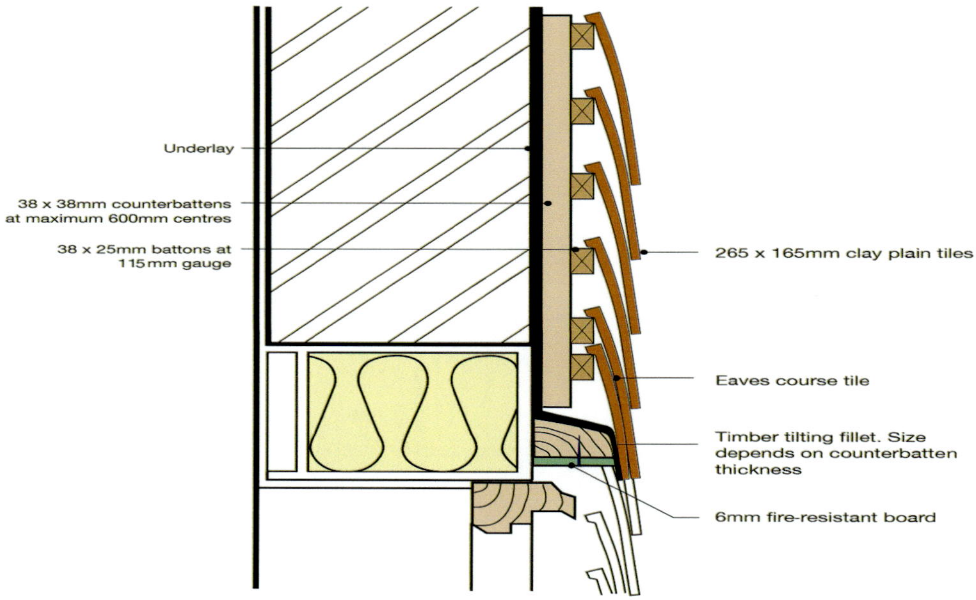

Figure 7.10 Vertical tiling to window head detail version 2

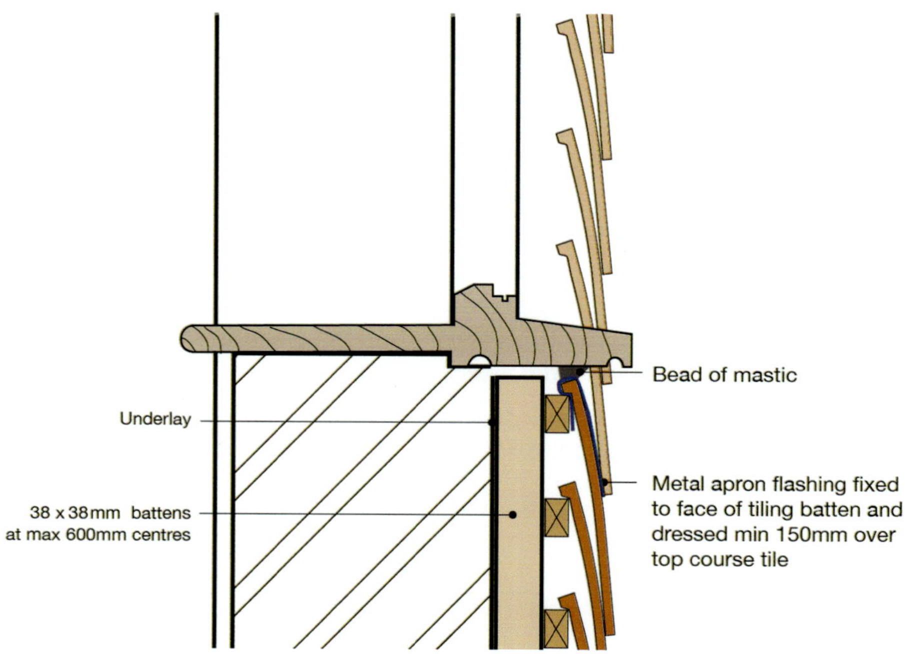

Figure 7.11 Vertical tiling to timber sill detail

Where there is a risk of fire entering the batten cavity, fix a minimum 6 mm thick fire-resistant board to the underside of the tilting fillet.

Abutment design details

The following illustrations and text explain the common main vertical tile design details that can occur on refurbishment and new work to abutments.

Vertical tiling junction with side abutment

Extend underlay on the main roof vertically up the wall by a minimum of 50 mm above the line of the tiling, and overlap by vertical underlay. The vertical upstands of the Code 3 lead soakers inserted between each course of roof tiles should be secured behind the battens/counter battens of the vertical tiling. Fix a timber tilt batten to the rake of the roof tiling to provide support for raking cut vertical eaves tiles. Cut tiles neatly and as close to the main roof tiling as possible.

Flashing to top abutment

Extend underlay on the main roof vertically up the wall by at least 50 mm above the line of the tiling, and overlap by vertical underlay.

Dress a Code 4 lead cover flashing over the top course of roof tiles and the tilting batten, carrying up under the underlay and secure behind the bottom vertical tiling batten. The eaves course of vertical tiling should be positioned closely to the top course of roof tiles.

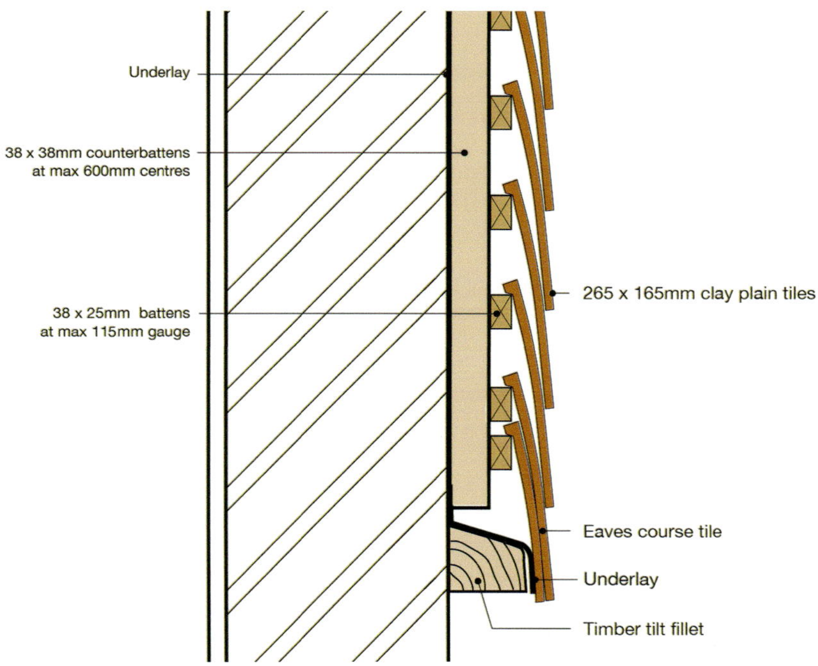

Figure 7.12 Vertical tiling eaves detail

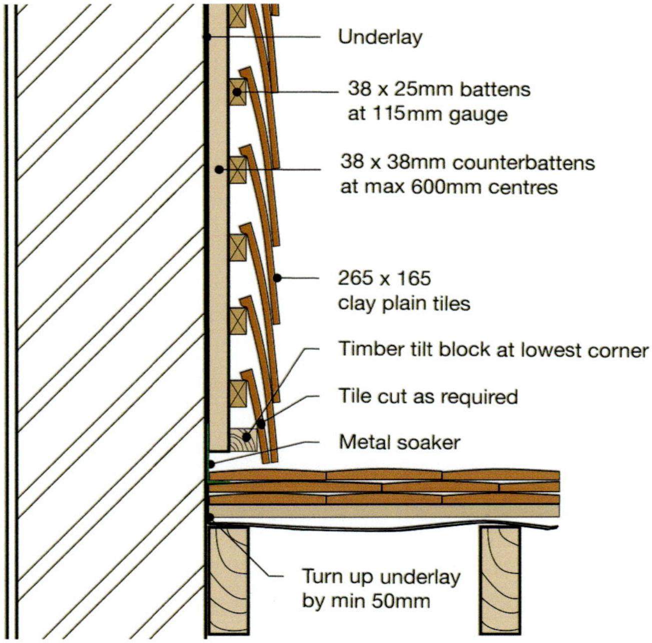

Figure 7.13 Vertical tiling side abutment detail

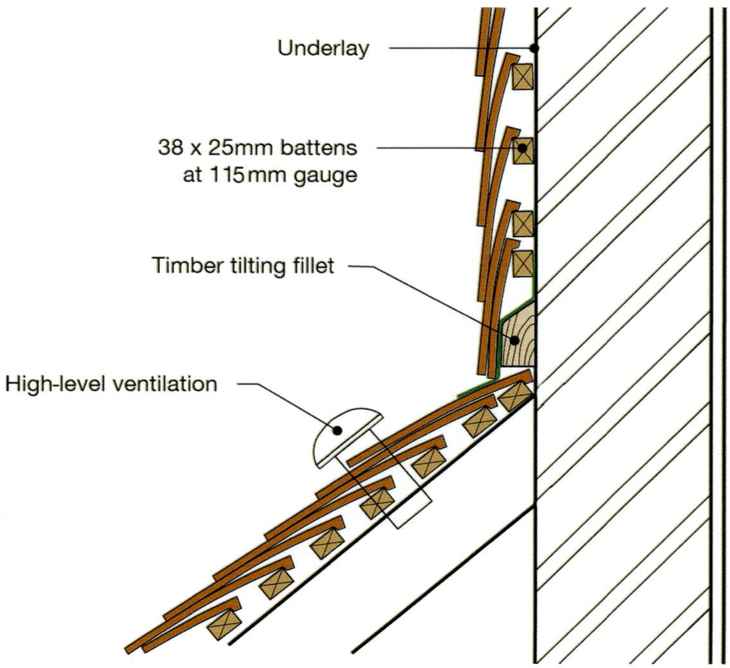

Figure 7.14 Flashing top abutment detail

Ventilated flashing to top abutment

Cut back the underlay on the main roof to provide at least 10 mm gap at the junction with the vertical wall. Dress a Code 4 lead cover flashing over the proprietary abutment ventilator and the timber tilting batten, carrying up under the underlay and secure behind the bottom vertical tiling batten. Position the eaves course of vertical tiling closely to the top course of roof tiles and ensure that a minimum 5000 mm^2 per metre run air path is maintained.

Flashing to mansard roof

Fix continuous timber tilt batten at eaves to provide support for Code 4 lead cover flashing and an eaves course of tiles. Ensure that a clear ventilation path is maintained from eaves to the ridge where insulation is positioned between the rafters.

Lap roof underlay over the lead welt and secure mansard underlay under tilt batten. Dress cover flashing over the top course mansard tiles by minimum of 150 mm and extending 150 mm to 200 mm up the tilt batten. Clip the bottom edge of the flashing in exposed locations.

Mansard roof with mansard tiles

Ensure that a clear ventilation path is maintained from eaves to ridge where insulation is positioned between the rafters. Lap the roof underlay over the mansard underlay by a minimum of 150 mm and establish the correct number of courses of mansard tiles to maintain a minimum head-lap of 65 mm.

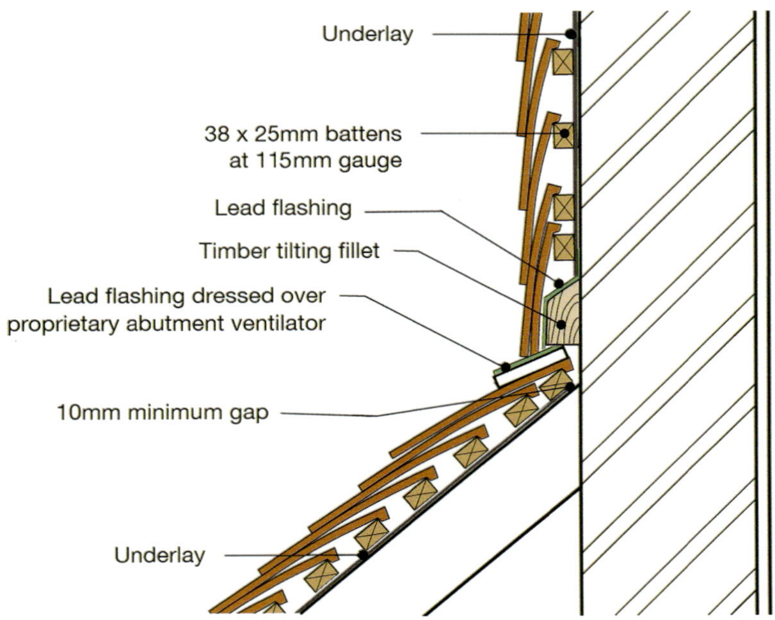

Figure 7.15 Ventilated flashing top abutment detail

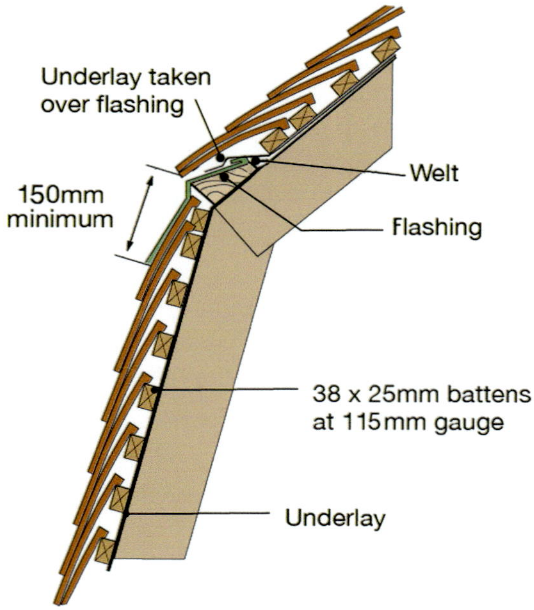

Figure 7.16 Mansard flashing

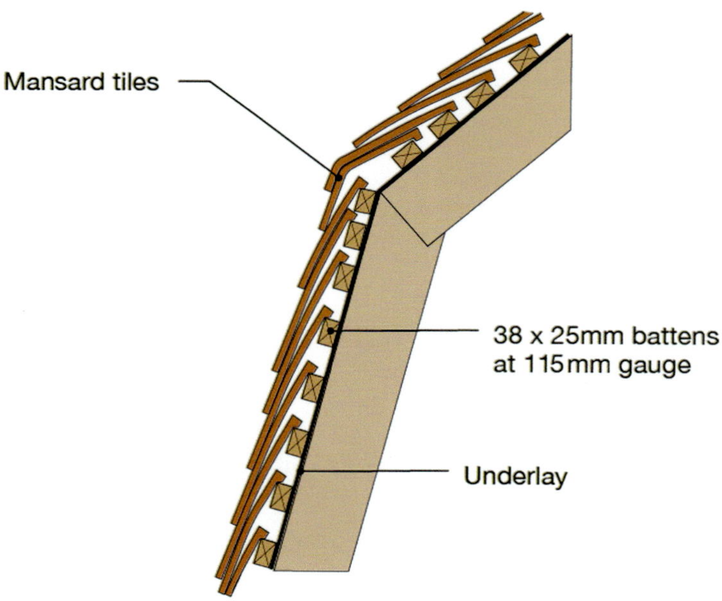

Figure 7.17 Mansard tiles

Vertical tiling junction with monopitch roof

Provide continuous timber fascia located directly above top course vertical tiles and fix Code 4 lead cover flashing to fascia and cover with monoridge tiles.

Dress lead cover flashing over the top course tiles by at least 100 mm. Install vent tiles spaced to provide the equivalent of 5000 mm^2 per metre high level ventilation to the main roof space.

Vertical tile design details to windows

The following illustrations and text explain common vertical tile design details that can occur on refurbishment and new work to window frames.

Vertical tiling to frame with pointed verge finish

Where the window frame is set back from the wall line and is installed from the inside of the building, the tiling traditionally has been finished flush with the window opening and then the window opening reveals were rendered to cover the ends of the vertical tiles. The distance from the face of the windows to the face of the tiling battens is not critical.

The tile battens should be cut approximately 25–50 mm short of the end of the tiling and treated to protect the batten ends from rotting. This detail works better where there is a short sill to the window that starts below the eaves line of the vertical tiling, such as French windows.

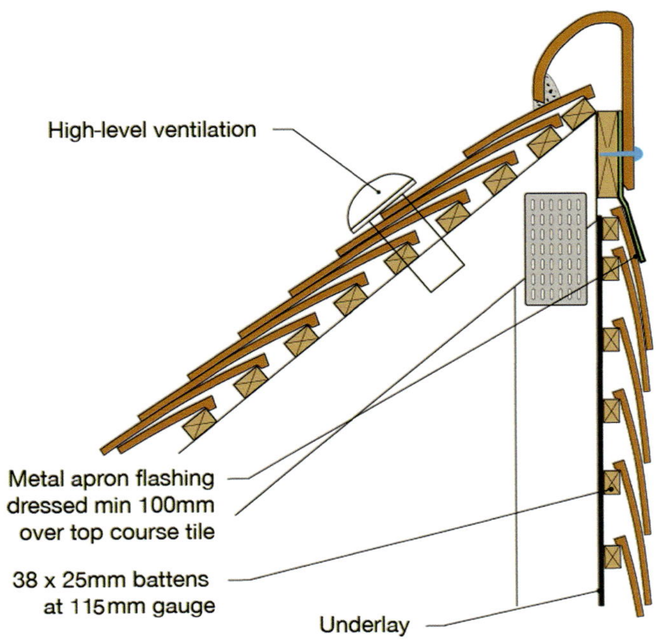

Figure 7.18 Monopitch detail

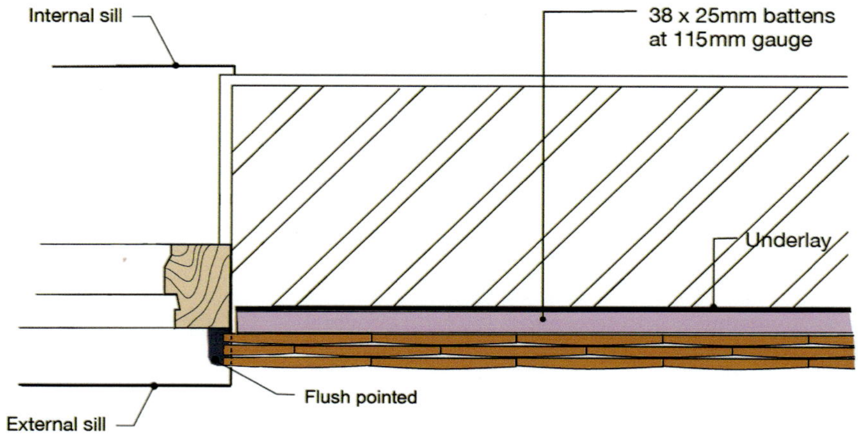

Figure 7.19 Pointed verge finish – plan view

Vertical tiling to inset frame

Figure 7.20 shows two versions of using external angle tiles to finish into the window frame.

In version 1, which works best where counter battens are not used, the window position should be set to eliminate or minimise tile cutting. Tiling should finish with handed external angle tiles cut every alternate course to ft the frame. The cut angle tile will require an additional fixing to maintain security. The joint between the tiles and the window frame should be bedded in mastic.

In version 2, the angle tiles are not cut into the frame (unless absolutely necessary), so the distance of the frame to the face of the tile batten is set to 125 mm. If this is not possible, then the frame is set to minimise tiles of less than half size width. A vertical flashing approximately 200 mm wide gives added weather tightness.

Corner tile design details

The following illustrations and text explain the main common tile design details that can occur on refurbishment and new work with internal and external corners joining tile to tile.

External angles with angle tiles

The first counter batten should be fixed approximately 20 mm in front of the end of the batten to prevent the end nail from splitting the tile batten. The cut ends of the tile battens should be alternated to coincide with the short leg of the external angle tile.

Handed external angle tiles are laid alternately up an external corner and each external angle tile should be twice nailed. Avoid cutting the external angle tiles to make them fit.

External angles with metal soakers

The first counter batten should be fixed approximately 20 mm in front of the end of the batten to prevent the end nail from splitting the tile batten.

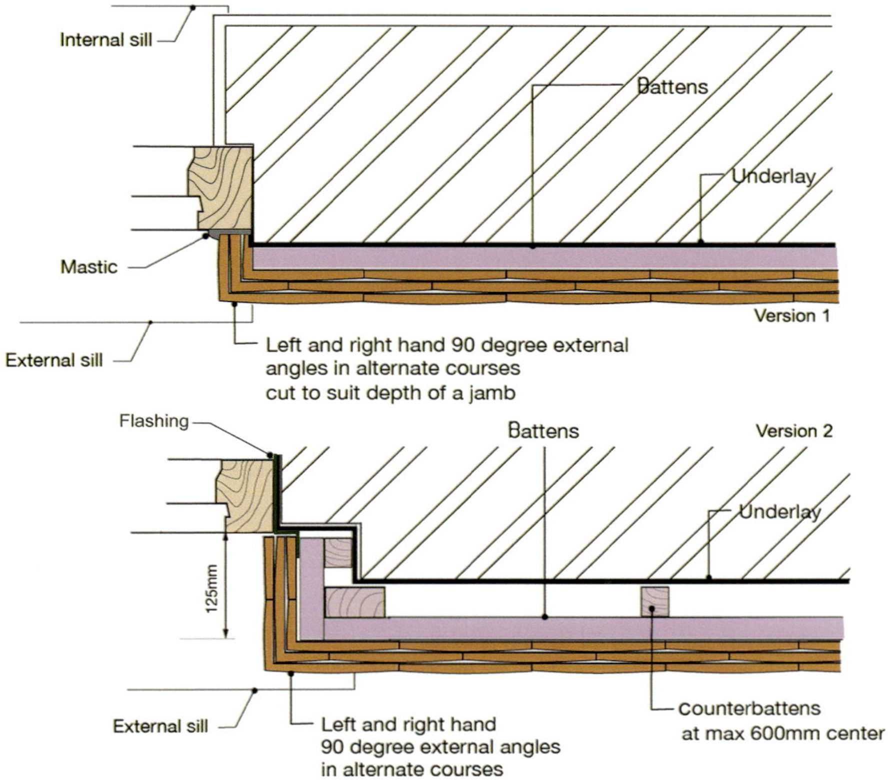

Figure 7.20 Inset frame finish – plan view

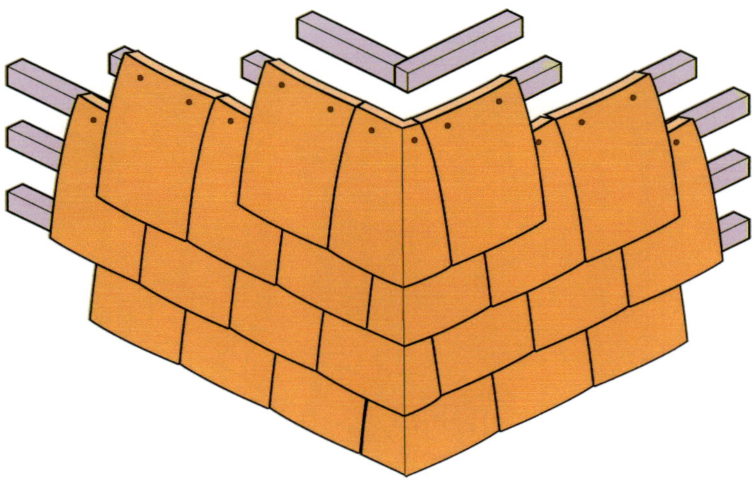

Figure 7.21 External angle tiles

In this situation, a Code 3 lead soaker or equivalent is required on each course of tiles. The soaker is cut and folded from sheet a minimum of 200 mm wide and 200 mm long. The bottom edge of the soaker should be flush with the bottom edge of the tile course above and the top folded over the head of the tile.

Use a combination of full tiles and tile-and-a-half tiles cut to a mitre in alternate courses, to maintain the bond. All tiles should be nailed twice.

Internal angles with angle tiles

The first counter batten should be fixed approximately 20 mm in front of the end of the batten to prevent the end nail from splitting the tile batten. The cut ends of the tile battens should be alternated to coincide with the short leg of the external angle tile.

Handed internal angle tiles are laid alternately up an internal corner. Each internal angle tile should be twice nailed and avoid cutting the internal angle tiles to make them fit.

Internal angles with metal soakers

The first counter batten should be fixed approximately 20 mm in front of the end of the batten to prevent the end nail from splitting the tile batten. In this fitment, a Code 3 lead soaker or equivalent is required on each course of tiles. The soaker is cut and folded from sheet a minimum of 200 mm wide and 200 mm long.

The bottom edge of the soaker is flush with the bottom edge of the tile course above and the top is folded over the head of the tile. Use full tiles and tile-and-a-half tiles cut to a mitre in alternate courses to maintain the bond and twice nail all tiles.

Flashing to side abutment with soakers

Finish battens approximately a 10 mm short of abutment wall and turn the underlay around the internal corner and under the flashings by 50 mm.

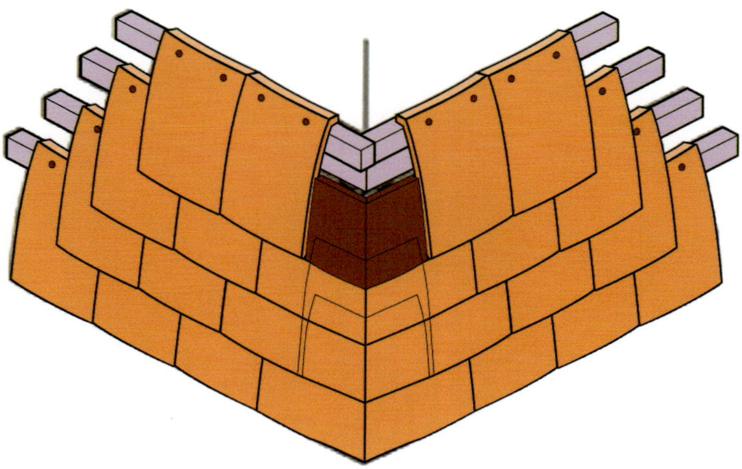

Figure 7.22 Metal soaker angles

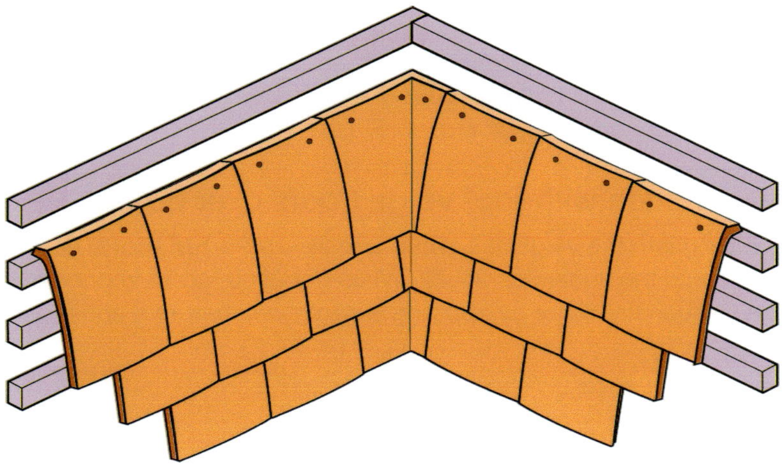

Figure 7.23 Internal angle tiles

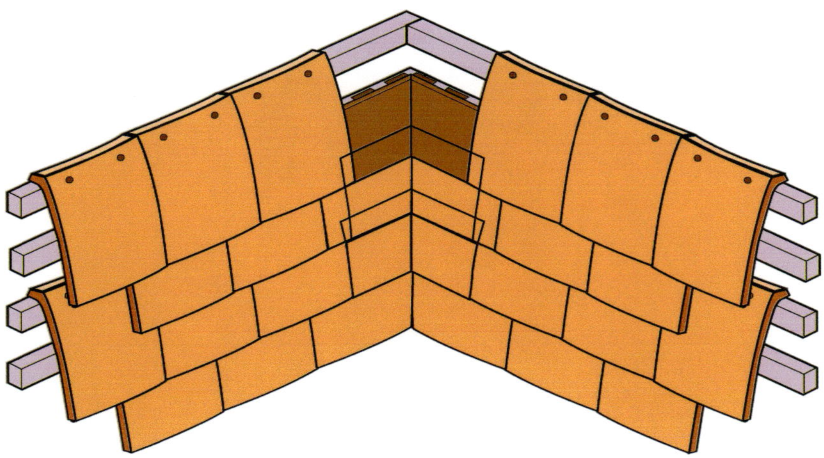

Figure 7.24 Internal angle soakers

Fix the first counter batten approximately 20 mm in from the end of the batten to prevent the end nail fixing from splitting the tile batten.

Fix a soaker, cut and folded from Code 3 lead sheet or equivalent, at least 200 mm wide and 200 mm long, to each course of tiles.

The bottom of the soaker should be flush with the bottom edge of the tile in the course above and the top should be folded over the head of the tile to prevent it dropping out. Turn a cut and folded strip of Code 4 lead, which should be 1.5 m long and approximately 100 mm wide, into the joints in the brick or blockwork, covering the exposed ends of the soakers by approximately 50 mm.

Detailing against a roof verge

The following illustrations and text explain the common cutting design details that can be used when finishing vertical tiling against a roof verge. The two most popular designs are the traditional Winchester cut and Sussex cut.

Vertical tiling junction with roof verge (roofs over 60°)

In general, when tiling into the verge at a gable, use tile-and-a-half tiles in each course of tiling, neatly cut to fit close to the undercloak/soffit. Where required, add a second mechanical fixing by drilling a hole through the tile or by applying suitable weatherproof adhesive.

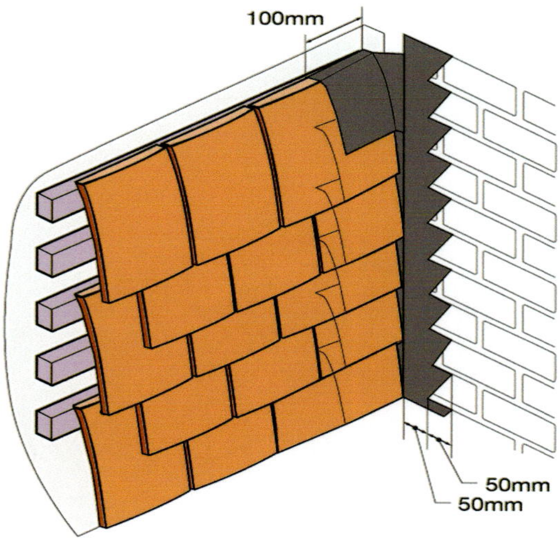

Figure 7.25 Side abutment soakers

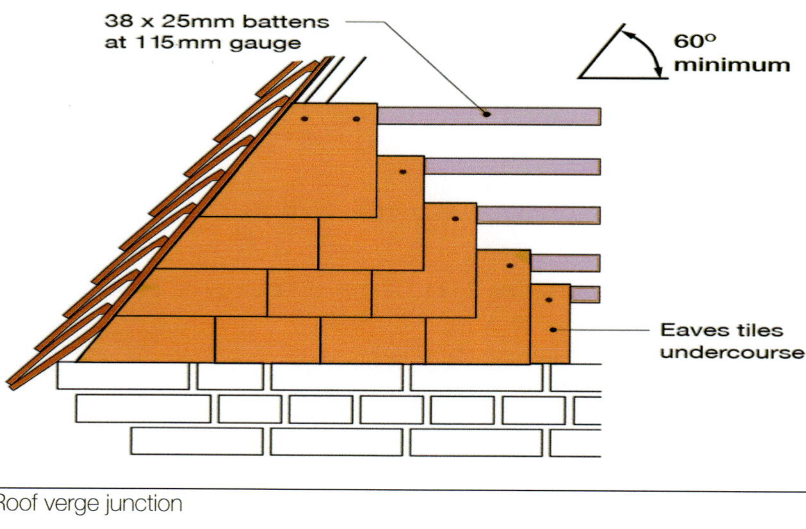

Figure 7.26 Roof verge junction

Vertical tiling junction with roof verge – Winchester cutting

A Winchester cut is a traditional finish for vertical tiling against a verge. Its distinctive look is the formation of continuous lines along each tile course. The finish at the apex is best achieved using a shaped piece of tile which is both mechanically fixed and spot bedded.

Winchester tiling is best suited for gable ends where the roof pitch is 40° or more. Between 35° – 40° where the side-lap may be below 55 mm the use of 200 mm x 200 mm soakers will be required.

Figure 7.28 shows three typical variations of Winchester cuts based on the roof pitch.

Vertical tiling junction with roof verge – Sussex cutting

Fix an additional batten onto the face of the vertical tiling battens and parallel to the verge to allow fixing of cut tiles.

Form raking cuts using tile-and-a-half tiles as necessary. Fix the edge tile close to the undercloak/soffit, securing by weatherproof adhesive and double nailing into the raking batten.

Soldier course

Vertical tiling beneath verges less than 40° roof pitch can be weathered by nailing a soldier course of eaves tiles to a raking batten fixed close to the undercloak/soffit. The raking batten should be positioned to allow the tiles to hang properly on their nibs, this may mean turning the batten 'on edge'. The use of weatherproof adhesive is recommended to prevent movement by wind and consequential damage. A cover flashing can be used to provide extra weathering over the soldier course tiles' nail holes and to the apex.

Figure 7.27 Winchester cutting

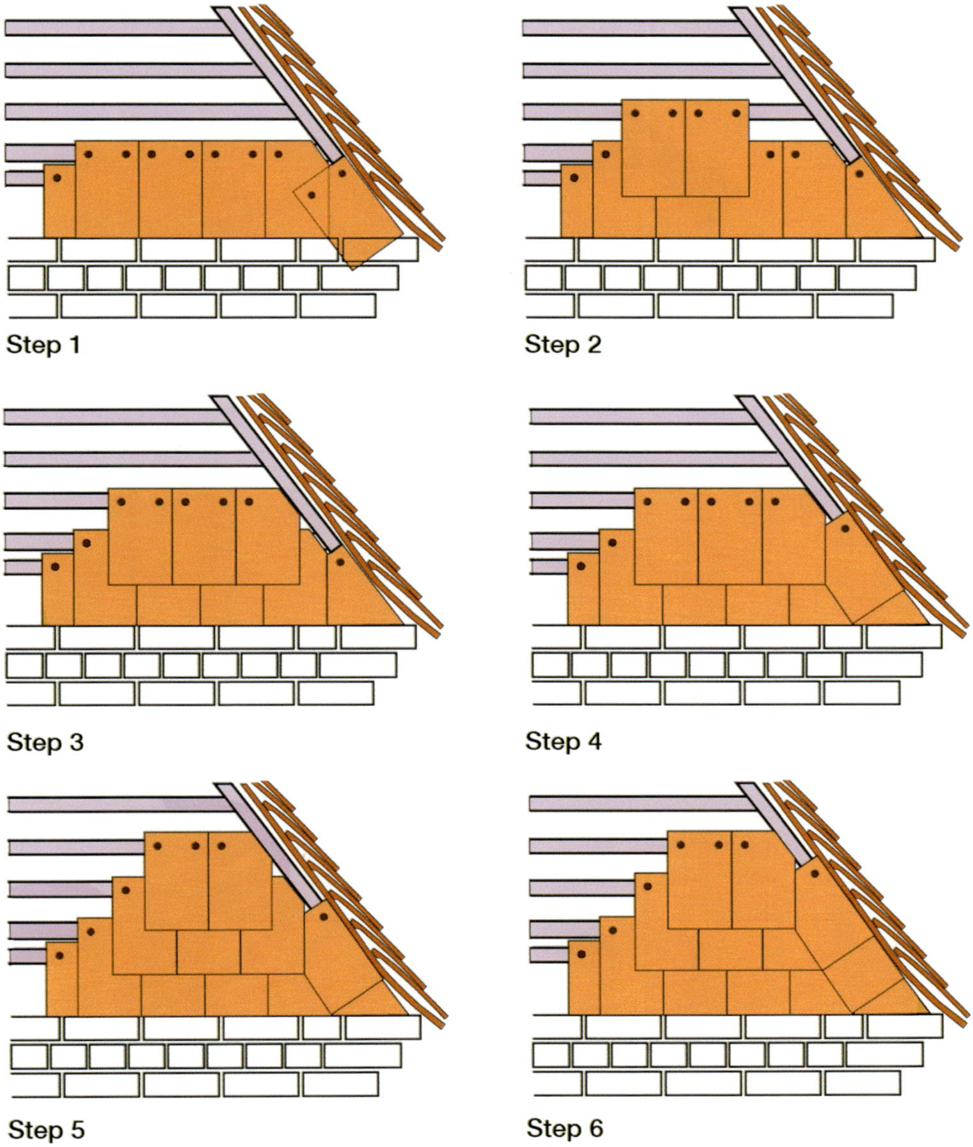

Figure 7.30 Sussex cutting stages

Double soldier course

With double soldier tiling, the method of fitting the tiles is the same as single soldier coursing, but the extra weatherproofing is obtained by using additional course of tiles rather than a cover flashing.

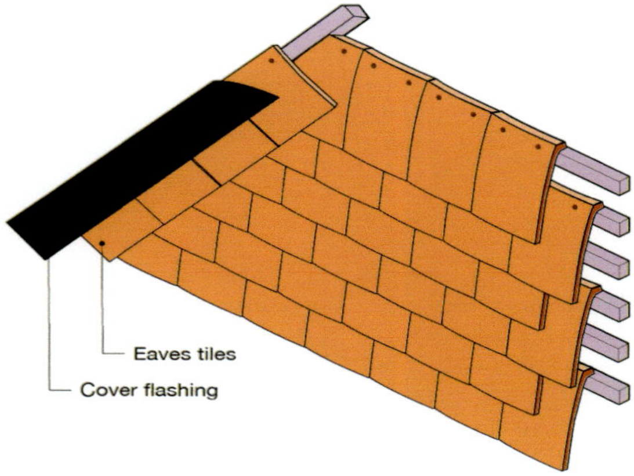

Figure 7.31 Soldier course

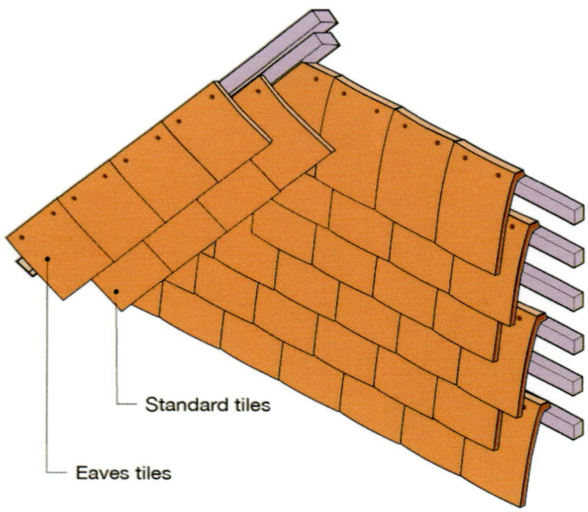

Figure 7.32 Double soldier course

7.4 Specialist techniques

Rules for side-lap and head-lap

It is important, particularly with the smaller sized tiles, to ensure that the rules for side-lap and head-lap are remembered and used.

The side-lap between tiles in subsequent courses must not be less than one third of the width of a standard tile, therefore, for plain tiles measuring 265 mm x 165 mm, the side-lap must not be less than 55 mm.

The head-lap between plain tiles in the course but one below must be not less than 65 mm.

Complex hip roofs

As well as standard hip roofs, there are a number of additional types that are rarer and sufficiently complex to warrant explanation of their specialist tile laying techniques. The following sections highlight the most common types of specialist hip roofs.

Hexagonal and octagonal towers with hip tiles

Arris hip tiles for Hexagonal and Octagonal Towers will normally need to be made bespoke to special order to suit the pitch of the tower. Standard sizes available are 120 and 135 degrees, referring to the angle on plan, as compared to a normal 90-degree corner.

At each hip, left- and right-handed arris hip tiles are used in alternate courses to maintain a broken bond to each side of the hip.

Octagonal (135°) tower with mitred hips

Tiles and tile-and-a-half tiles are cut and fixed to form a straight, weathertight, close mitred junction at the hip. The mitred tiles are interleaved with metal soakers, extending a minimum of 100 mm to each side of hip. The soakers are secured by turning down over the heads of mitred tiles. Extreme care is needed to achieve a neat finish at the hip. Where possible, it is advisable to use specially made arris hip tiles to suit the particular roof pitch instead.

Mitred hips

Lay courses of underlay over hip with overlaps of not less than 150 mm. Cut tile-and-a-half tiles and fix to form a straight, weathertight, close mitred junction. Interleave mitred tiles with metal soakers, extending a minimum 100 mm to each side of hip, and fix soakers by turning down over heads of mitred tiles. Extreme care is needed to achieve a neat finish at the hip. Where possible, it is advisable to use either bonnet hip tiles or specify specially made arris hip tiles to suit the particular roof pitch instead.

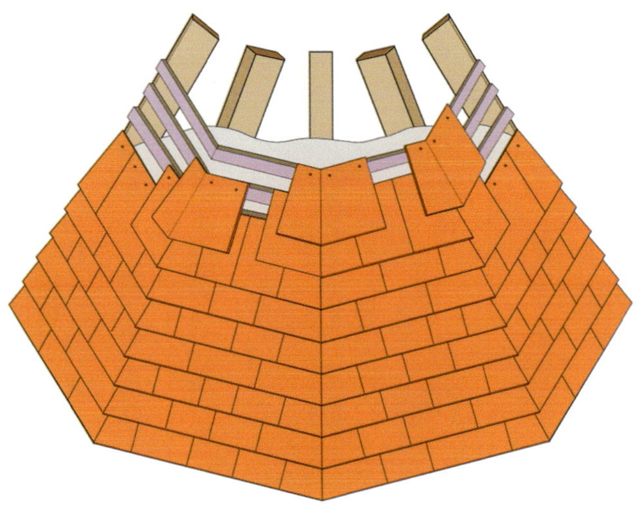

Figure 7.33 Handed arris hip tiles

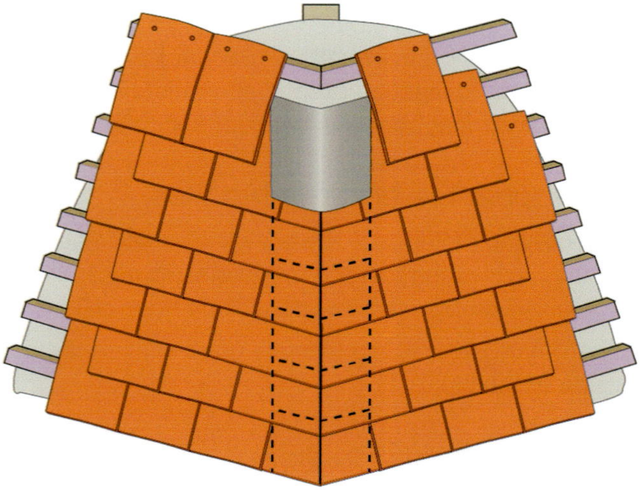

Figure 7.34 Mitred hip detail used on an octagonal

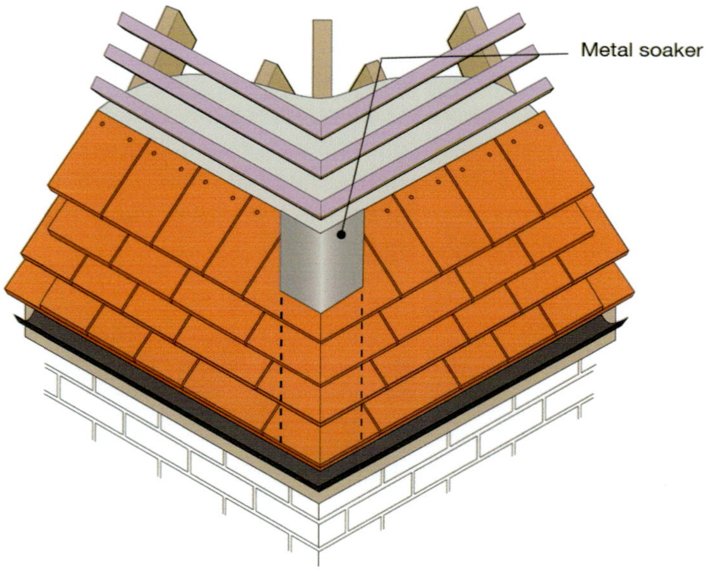

Figure 7.35 Mitred hips

Four-way hip cap

Where four mitred, bonnet or arris hips terminate without a ridge line, a purpose-made cap or finial can be used. Because hip caps and finials are pitch specific, please contact the tile manufacturer with roof details. If needed, decorative finials can be made to order.

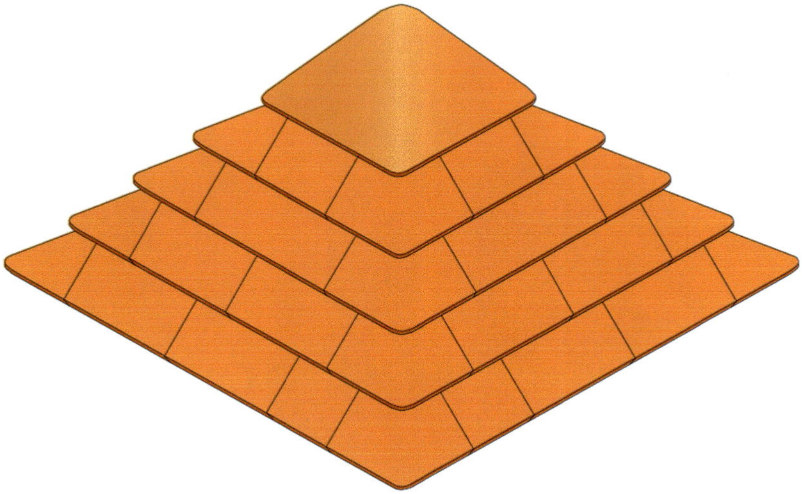

Figure 7.36 Four-way hip cap

Hips with unequal roof pitches

If it is required to continue the tile courses in line around the hip line, the roof pitches should be equal to achieve the best result. However, when the roof pitches are unequal (up to 5 degrees) it will be necessary to set out the tile batten gauge on the lower pitch first. Then, the battens on the steeper pitch can be fixed in line with the first battens. This will result in a shorter gauge on the steeper side.

Weatherproofing with soakers

Traditionally made of lead, and used with double-lap tiles, soakers are placed in between tiles at junctions where there is a high risk of water ingress, in order to minimise the risk and enhance durability. Modern popular alternatives to lead are aluminium and polypropylene, and these normally are preformed.

Side abutment

Soakers can be cut from Code 3 lead or equivalent. The length of each soaker should equal the tile gauge, plus the head-lap, plus an extra 25 mm to turn over the top of the tile. For example, 100 mm + 65 mm + 25 mm = 190 mm.

The width of the soakers should be 175 mm to provide 100 mm lap onto the tile and 75 mm upstand at the wall. The step cover flashing must cover the soakers by at least 65 mm.

Junction of ridge with abutment

The lead saddle at an abutment should extend 150 mm along the ridge and 150 mm down each roof slope. The edge of the saddle under the ridge tile should terminate in a welt (a folded edge that strengthens the flashing) and for steeper roof pitches it may be necessary to form the saddle using lead welding.

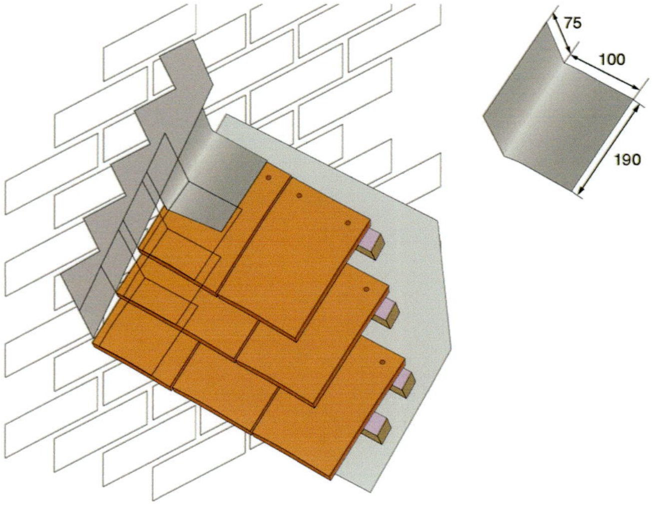

Figure 7.37 Side abutment soaker detail

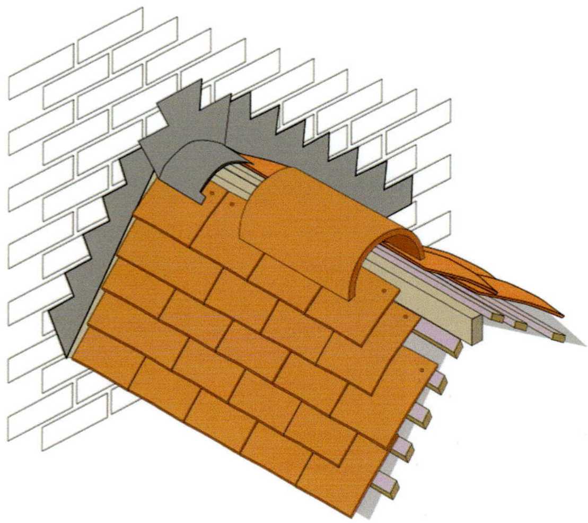

Figure 7.38 Ridge abutment soaker detail

Junction of ridge with hip

At a junction of a ridge with a hip, a lead saddle should be fitted. This should extend at least 100 mm down each roof slope and the edge of the saddle under the ridge tile should terminate in a welt.

Figure 7.39 Ridge and hip soaker detail

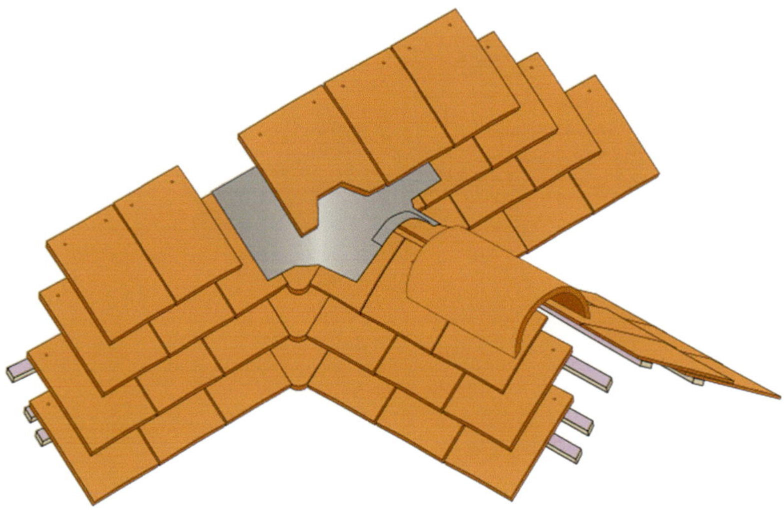

Figure 7.40 Ridge and valley soaker detail

Junction of ridge with valley

A lead saddle should be fitted where two tiled valleys meet. This should extend at least 100 mm down each roof slope. The top edge of the saddle should extend at least one tile headlap and terminate by turning over the head of the tile, and the edge of the saddle under the ridge tiles should terminate in a welt.

Mitred valley

Valley soakers should be used on every tile course in a mitred valley detail. Each soaker should extend at least 150 mm to either side of the valley and be held in position by turning over the heads of the tiles below.

Valley soakers are not recommended for use on roof pitches below 50°, on valleys longer than 6 m, or where water is discharged into the valley from other roof slopes.

Mitred hip

Hip soakers should be used on every tile course in a mitred hip detail. Each soaker should extend at least 100 mm to either side of the hip and is held in position by turning over the heads of the tiles below. Hip soakers are suitable for all plain-tile roof pitches.

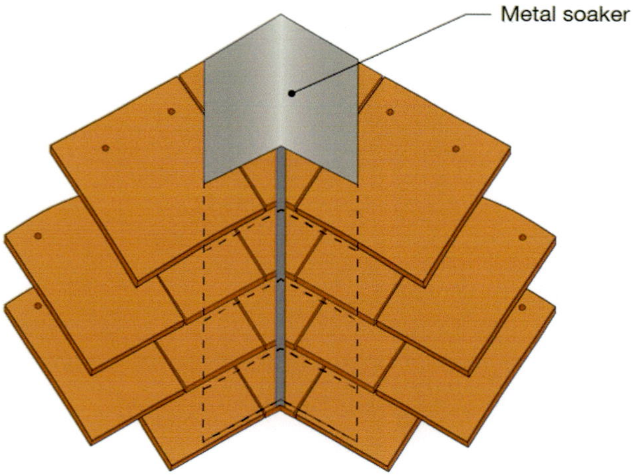

Figure 7.41 Mitred valley soaker detail

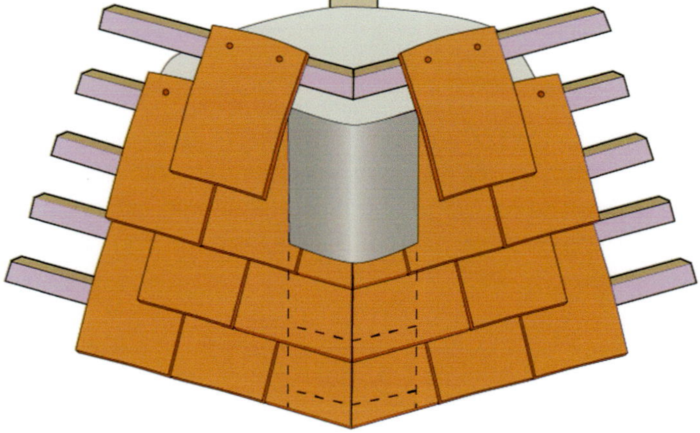

Figure 7.42 Mitred hip soaker detail

7.5 Secret box gutters

The secret box gutter runs horizontally across the roof, usually about 300 mm above the eaves. The gutter may be constructed in timber and usually lined with lead or other suitable material. It can also be a preformed system made from aluminium or PVC extrusion.

The ridge side of the gutter can be treated as eaves and is finished with an eaves tile course, projecting 50 mm over the gutter. The tails of the eaves course tiles should be in line with the tails of the first course of full tiles above the gutter.

The lower side of the gutter is treated as a top course and will have a short tile course over the last full tile course. The lead lining of the gutter should overlap the top course tiles by at least 100 mm. This detail can be used to break up long rafter lengths. The following diagram in Figure 7.43 shows an example of a gutter positioned a short distance up the roof slope from the eaves.

7.6 Bell-cast/sprocketed eaves

Bell-cast eaves were common in rural Dutch colonial architecture. They can be defined by a gradually diminishing slope, which usually flares out to create an eave that projects well beyond the face of the supporting wall.

Because the minimum recommended roof pitch for plain tiles is 35° (40° for some handmade plain tiles), the general roof pitch needs to be greater to ensure that the bell-cast is at least the minimum pitch. Where the minimum head-lap cannot be maintained, soakers should be used to reduce the risk of water penetration.

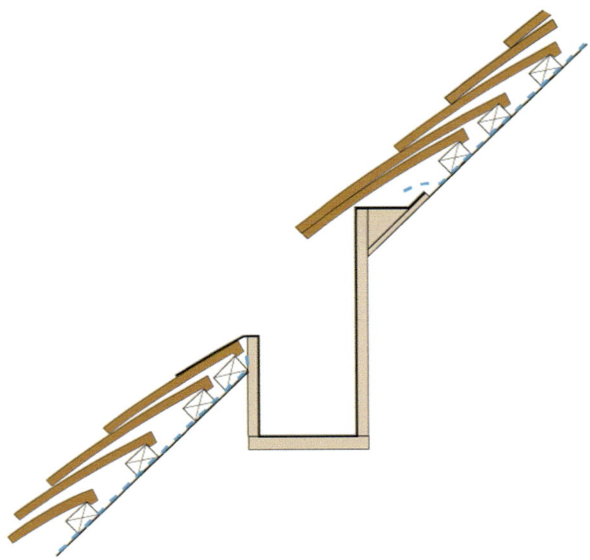

Figure 7.43 Secret box gutter section

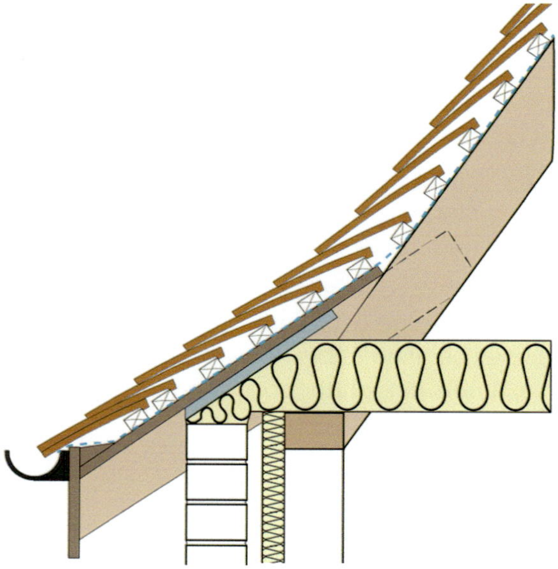

Figure 7.44 Section through eaves showing 'bell-cast' or 'sprocketed' eaves

7.7 Eyebrows

An eyebrow dormer, sometimes known as a roof eyebrow, is a wavy dormer that protrudes through the slope of a roof. It usually contains a window that may be fixed or operable.

Construction is difficult because there is not a defined junction and the framing and finishing can be tricky.

Bending and scarfing of battens

To enable tile battens to be bent over the curve of the eyebrow they are 'scarfed' by sawing a series of short cuts into one face of the batten, and this allows the batten to bend in the scarfed direction. However, scarfing with modern 25 mm thick battens is very difficult on all but very gentle curves so it is normally better to use built up layers of 6 mm battens for any curved work. For example, four layers of 6 mm battens could be used.

If practical, and if there are facilities on site, soaking and/or steaming the battens will greatly aid their ability to bend and conform to the curved roof surface.

Critical dimensions

The minimum general roof pitch should be 55° to ensure that the eyebrow minimum pitch is 35°. Where the roof pitch is less than 55°, soakers should be used with the tiles on the eyebrow. The slope of the curve should be the minimum possible to avoid the risk of tile 'chatter' which is the sound caused by prevailing winds lifting and dropping

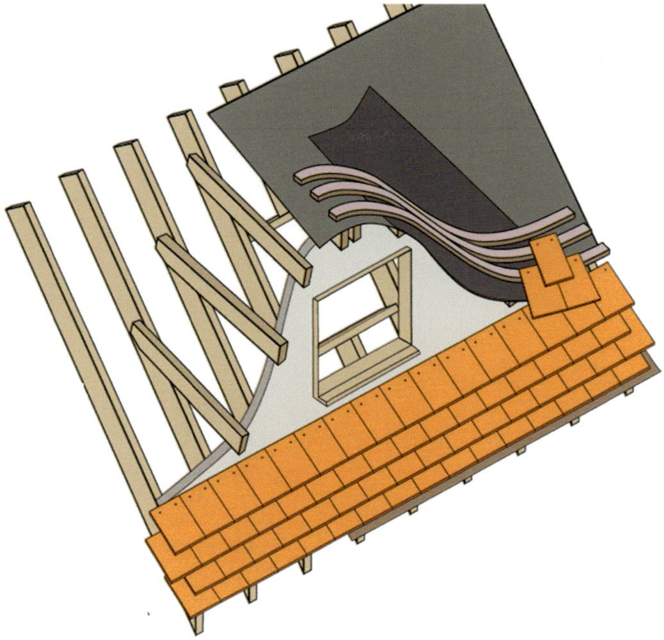

Figure 7.45 Eyebrow construction

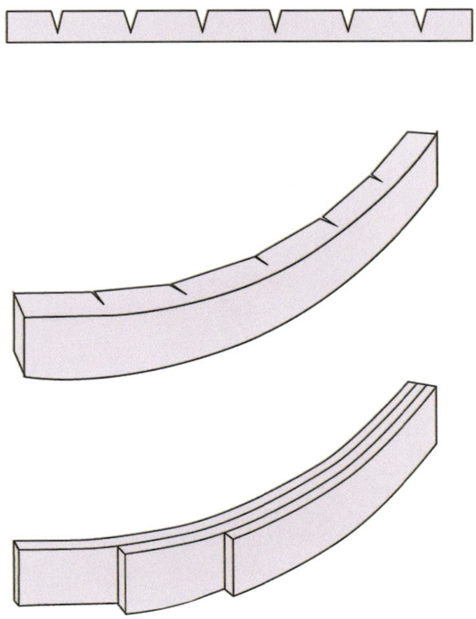

Figure 7.46 Bending and scarfing

slightly the bottom edge of the tiles. Tile chatter can occur on steep curves where the tiles do not fit closely round the top of the curve.

The transition between eyebrow and general roof should be regarded as a valley, and precautions such as courses of extra underlay, and strip soakers should be taken to eliminate the risk of water ingress. To further avoid the risk of water ingress, the eyebrow should start/terminate no closer than 600 mm (but preferably 900 mm) from roof edges such as hips, valleys, abutments, etc.

The apex of the curve of the eyebrow should occur between rafters to ensure a smooth curve to the tile battens and rafter centres should not exceed 400 mm.

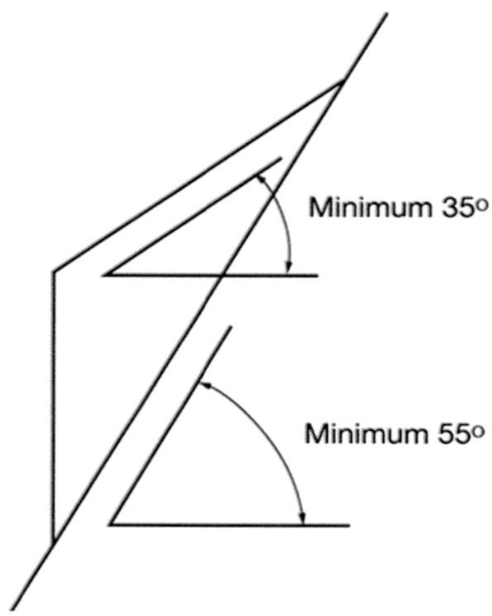

Figure 7.47 Critical pitch dimensions

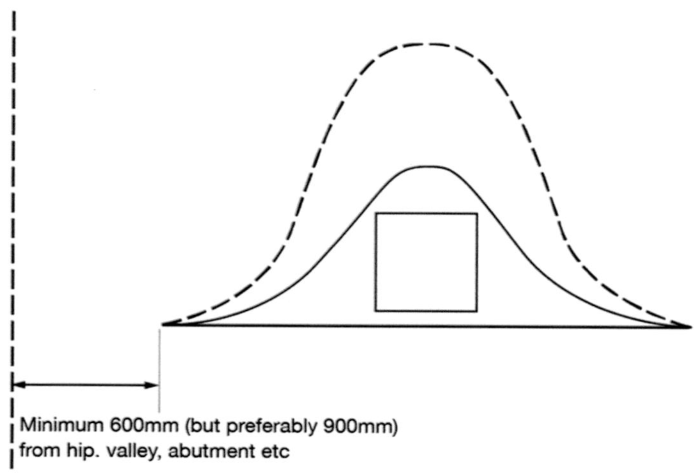

Figure 7.48 Interface with main roof

Head-lap and side-lap

During the eyebrow construction, it is important to ensure the tiles keep their side-lap and head-lap throughout.

7.8 Domes and cones

Domes and their substructure

Dome tiles, or 'fish-scale' tiles, should be used. These are semicircular at the tail and square at the top, with the sides tapered to absorb the horizontal curvature. The tiles should have little or no camber. The size of the tiles in relation to the size of the dome is small so there should be little risk of tiles 'riding' on each other. The shaping of the tails compromises weather efficiency but makes the tiles more flexible in use.

The pitch of the dome almost flattens out at the top of the roof so tiles here cannot provide any weather protection. Therefore, a complete covering of lead or other suitable metal is required to ensure adequate weather protection. Such a covering, or cap, can be decorative, or incorporate a window detail to allow light into the dome.

Cupola roofs

A cupola is a small dome, but at the critical roof pitch the dome rises to a peak to avoid the very low roof pitch angles. As the tiling reaches the apex, normally, the minimum side-lap cannot be achieved and hence soakers will be required to maintain the weather tightness.

The cupola groundwork can be a complex layer of 'plasterers laths' laid obliquely by bending the laths over all the curves to allow tile nails to be driven in anywhere on the curve.

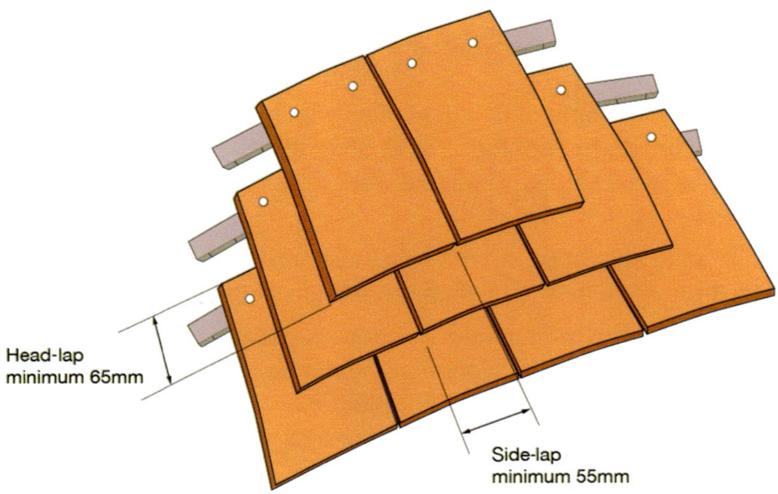

Figure 7.49 Minimum head-lap and side-lap

Figure 7.50 Tiled dome roof

Alternatively, 38 mm x 25 mm battens could be used, amply 'scarfed' or otherwise treated to allow bending to the curves.

Conical roof

A conical roof is a cone-shaped roof that is circular at its base and rises to finish in a smaller diameter.

As shown in Figure 7.52 below, to achieve a smooth curve across the rafters on a conical roof, the roof is usually boarded with exterior grade plywood before installing the battens and tiles. The layers of thin plywood are laid with staggered joints to create a depth of 25 mm.

The underlay should then be laid vertically over the plywood, with vertical laps of 150 mm and horizontal laps of 100 mm.
To comply with the requirements of BS 5534, tiles should be fixed to timber battens, but as scarfed battens or strips do not conform to BS 5534 this is 'out of scope' work. In practice, and as seen for heritage and other traditional work, tiles can be fixed directly to the plywood and suitable underlay.

The tile battens (if used) should be set out to a maximum of 100 mm for plain tiles, or 95mm for peg tiles, which provides for the minimum head-lap of 65 mm.

Each course of tiles going up the tapered structure is a decreasing circumference, and this will require the tiles to include some tapering of width so as to maintain the bond, or side-lap. For example, a typical conical roof of 70° pitch may require three sizes of tapered tile. The first of 150 mm width at the tail, the second of 125 mm and the third of 100 mm. To

Figure 7.51 Cupola roof

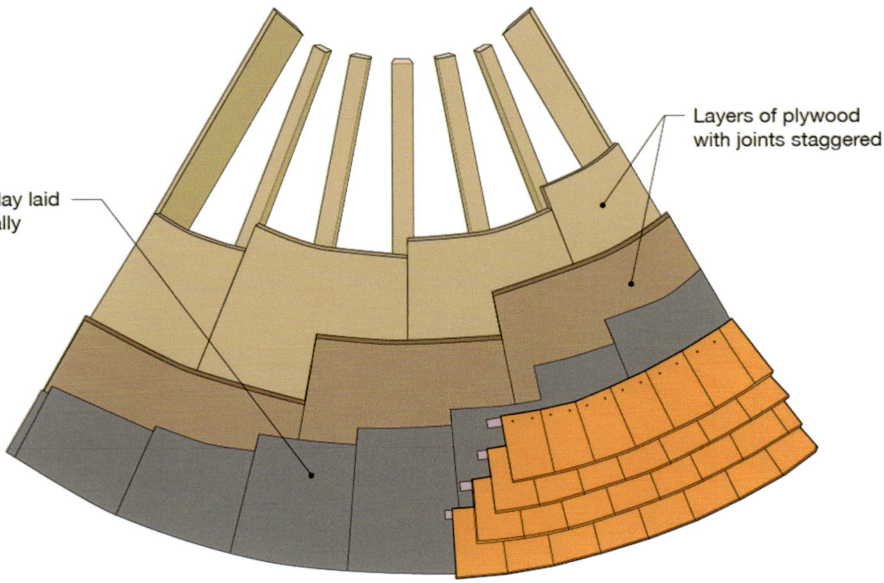

Figure 7.52 Plywood skin on conical roof

maintain an effective bond or side-lap with the adjacent courses, the minimum possible width of tile is 81 mm at the head.

Alternately, start with a felt template, manually cut tiles to one size per course until they get to 81mm at the head (½ a tile). Where 55mm side-lap cannot be maintained, install strip soakers between the courses.

Depending on the radius of the curve, it may not be possible to rely on any nibs for support, therefore each tile should be nailed. It is not good practice to drill two holes in an already thin cut tile, one nail hole plus adhesive would be considered more practical.

Purpose-made tapered peg tiles are available from some manufacturers for use on historical conical roofs, particularly oast houses.

In the top section of the cone, it is not practicable to make tiles narrow enough for the curvature, and still maintain sufficient side-lap, so therefore a large flashing cover/cap is used.

7.9 Convex and concave curves

Tiling a curved roof

The underlay must be laid vertically. The battens should be scarfed at the back to enable them to be bent around the curvature of the roof, or use built-up layers of 6 mm. Refer to Chapter 7.7 *Eyebrows* for more information on the shaping of battens.

As with conical roofs, each course is a different width and this will require the same methodology for tiling as does a conical roof.

It is important to remember that the success of this type of curved roof work very much depends on the skills of the roofing contractor. A curved roof is very labour intensive and therefore it is advisable to appoint a competent roofer who is experienced in this form of roofing.

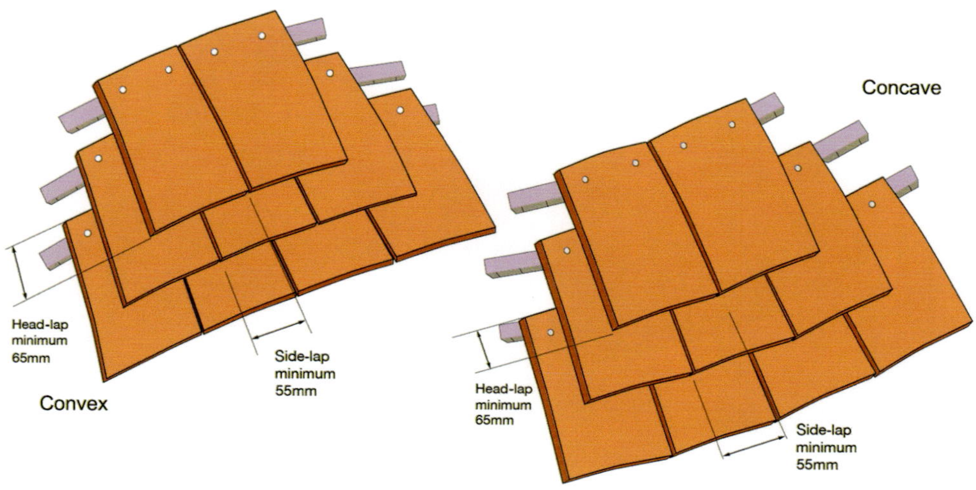

Figure 7.53 Convex and concave curved roof

CHAPTER 8

Checking and maintenance

8.1 Inspection requirements

The typical expected service life of clay and concrete roof tiles is between 30 to 60 years, but it is common knowledge within the industry that the lifespan of a roof tile can extend up to 100 years. Even though there have been improvements in material specification and quality control in the manufacturing process, proper maintenance of the completed roof is still required after installation.

To assist with this, manufacturers provide guidance on what to look for. The following inspection checklist highlights the areas that should be monitored:

- Missing, damaged or chipped tiles.
- Deterioration, creep or cracking of metal-lined valleys or fibre-glass valley troughs.
- Loose, broken or missing roof, ridge, hip and verge tiles.
- Disturbed metal flashings and cracked or dislodged mortar bedding.
- Eaves, gutters and guttering joints clear of debris.
- Excessive moss growth affecting the drainage channels and joints of tiles.

Maintenance

The main objective of regular maintenance involves carrying out regular visual inspections of the roof, usually twice a year, in the spring and autumn. Any leaves and debris should be removed from valleys and gutters, together with any moss or lichen growths that restrict the flow of water off the roof slope.

Check the condition and security of roof tiles, accessories and flashings, particularly at the more vulnerable perimeter areas of the roof at the ridge, hip, verge and valley and any abutments to walls or roof lights. Check the function of any roof space ventilation components and clear any grilles or apertures to ensure adequate airflow within the roof void.

Access equipment

The Work at Height Regulations require duty holders to ensure that all work at height is properly planned and organised, that those involved in work at height are competent, that the risks from working at height are assessed, and that appropriate work equipment is selected and used.

DOI: 10.1201/9781003196990-8

The risks associated with accessing fragile surfaces should be properly assessed, including the preparation of a risk assessment to identify the key hazards and a method statement confirming the actions required to address the identifiable risks to ensure safe systems of work compliant with health and safety regulations.

Any equipment used for work at height should be properly inspected and maintained prior to undertaking the work in accordance with regulatory guidance.

Ladders

Ladders should always be secured and should primarily be used for access only in order to undertake light work of short duration, and then only if it is safe to do so. It is generally safer to use a tower scaffold or a mobile elevated work platform (MEWP), even for short-term work. Heavy work activity carrying heavy loads should never be undertaken from a ladder. When using a ladder, ensure that the person on the ladder always maintains three points of contact, such as two legs and a hand.

Ladders should only be used for the purposes of inspection of the roof from the eaves or wall parapet or for simple maintenance tasks and should not be rested against the gutter or parapet. The ladder should be set against the vertical wall of the building at approximately 75° or the bottom of the ladder at a distance from the vertical plane equal to the total length of the ladder. The foot of the ladder should be sited on firm and level ground and made secure to a ground anchor or footplate.

The top of the ladder should be made secure to the structure using an appropriate ring bolt or hook, clamp or rope; should extend to project not less than 1.07 m above the eaves or parapet and be fitted with a wall stand-off to clear the gutter or parapet. All ladders should comply with the relevant standard BS EN 131.

Where access to the roof slope is required, proprietary roof ladders should be used in conjunction with appropriate working platforms. Care should be exercised when working near metal or open valleys so as not to damage the side coverings. GRP prefabricated valley trough units are vulnerable to breakage and should not be eased or levered or used for foot traffic.

All roofs clad with clay and concrete tiles and slates should be treated as fragile, and extra care must be taken if it is necessary to traffic them. Suitable packing material should always be provided between roof ladders and the actual covering material to prevent breakage, e.g. foam rubber.

It is not advisable to traffic roofs clad with proprietary lightweight resin-bonded tiles or slates and photovoltaic (PV) solar panels or integrated PV tiles, which may require special protection against damage. It is recommended that the manufacturer is consulted for guidance before gaining access to a roof clad with such products.

Safety hooks

Safety hooks are proprietary devices which are fixed directly to the roof or building structure and to which safety ropes or harnesses are attached by roofing or maintenance contractors. Such products are subject to health and safety legislation and should comply with BS EN 517 and BS 7883.

Roof walkways

Proprietary devices which are fixed to the roof structure or as part of the clay or concrete tile product may be used for access in order to stand or walk during inspection, maintenance or repairs to elements or parts of the building structure which penetrate the roof covering.

These devices may be required by health and safety legislation and should comply with BS EN 516.

Working platform

Where small areas of roofs are to be accessed for repair or maintenance, a working platform or mobile access platform may be required at eaves level. Mobile access platforms are only permitted in these circumstances.

All MEWPs should be constructed to the requirements of BS 7981 (power operated) or BS EN 1004 (mobile working towers) and BS EN 12810–1 (prefabricated facade scaffolds) The use of a MEWP, ladders or towers to access the roof requires special considerations.

Scaffold

If roofs are to be extensively repaired or reroofed, a working platform in the form of an independent tied scaffold conforming to BS EN 12811–1 should be constructed. A suitable guard rail or barrier must be provided at the edges of the roof (eaves, verges) where scaffold is not provided, and should be constructed in accordance with this standard.

Ensure that all roof inspections and roofing works are carried out by competent persons in accordance with health and safety regulations and guidance.

Repairs

All repairs, re-covering and maintenance of tiled roofs and walls should conform to:

a) current building and health and safety regulations and guidance
b) British Standards – BS 5534 and BS 8000–6.

Depending on the size of the repair or area to be inspected, access to the roof can either be temporary or permanent. Broken or defective tiles should be replaced and refixed with a sound matching unit and mechanical fixing as recommended by the manufacturer and not covered superficially with any other material or coating. If extensive repairs are required, sectional or complete re-covering should be considered. Proprietary surface coatings or spray foam roof undercoatings applied to weatherproof or insulate the complete roof externally or internally are not recommended.

Ridge and hip fittings should be replaced individually and refixed using recommended materials/fixings where required (see BS 5534, BS 8000–6 and manufacturer's literature).

8.2 Common maintenance issues

This chapter contains guidance to enable designers, contractors and building owners to identify common maintenance and repair issues that can affect roofs and walls clad with clay and concrete tiles and to implement appropriate design and maintenance measures.

It is acknowledged that all roofs require some level of maintenance during their lifetime, even if it merely involves the removal of windblown debris from valleys and gutters. Access to a roof is sometimes required in order to maintain other building elements, such as windows and chimneys. Consequential damage to the roof or wall cladding can also compromise the integrity of the roof covering, leading to water ingress and loss of safety of the roof element.

Before commencing to replace the roof covering or carry out any structural alterations to the roof, the designer or installer should refer to the current building regulations and their requirements.

When correctly installed in accordance with the recommendations of BS 5534 and BS 8000–6, a completed roof or wall clad with clay and concrete tiles should give trouble-free performance for the guaranteed life of the product without the need for extensive maintenance or repair.

To achieve the full benefits of a roof or vertical tiling, there are a number of standard procedures that should be drawn to the attention of the building owner or maintenance operative when occupying the completed building.

All tiled roofs and walls should be treated as fragile and basic precautions should be taken to avoid access to the roof, such as by window cleaners, chimney sweeps and aerial installers, without the correct use of crawling boards, roof ladders or access platforms. Failure to use adequate access equipment can damage the tiles and fixings and invalidate any guarantee, plus may be in contravention of health and safety regulations.

Maintenance items

Frost

Frost failures of mortar are mainly confined to low-strength mortars held in saturated conditions. Low-strength mortars subjected to soluble salts are also vulnerable.

Freshly laid mortar used with roofing fittings which exercise little suction is particularly unstable, but so is fully hardened mortar containing appreciably less than the prescribed amount of cement. The risk of frost failure of any mortar is considerably reduced by air entrainment.

Efflorescence and staining

Efflorescence is a general term used in the construction industry to describe the white deposit of calcium carbonate found on all concrete building materials. It commonly appears in the form of white patches or as a more general lightening in colour on the surface of concrete roof tiles and fittings. When the latter effect is seen, it is often misinterpreted as a fading or 'washing out' of the colour of the concrete.

Efflorescence forms more readily when the concrete tile becomes wet and dries slowly and therefore there are more occurrences during the winter period. However, it is only likely to occur in the early life of concrete roof tiles and fittings, and those installed for a year or longer that do not show signs of efflorescence are unlikely to be affected in the future.

Perhaps the most important factor for the specifier, builder and property owner is that the natural weathering process gradually removes the efflorescence. This natural removal also restores the original colour of the product and in no way affects the product's permeability or continuing strength growth with age.

Efflorescence can be considered a superficial characteristic of quality concrete roofing products. Some concrete tile manufacturers have developed surface coatings to help suppress the formation of efflorescence on the top surface of their tiles and fittings.

Repair procedures

Underlay

Repair any tears, holes or cuts in the underlay by cutting a slit above the hole and placing a sizeable piece of similar material large enough to fit under and lap onto the underlay around the hole by at least 150 mm.

Secure the replacement underlay material under the battens or fix to the adjacent rafters using nails or fixings as recommended by the manufacturer.

Battens

Defective battens should be replaced for a minimum of two rafter spacings to ensure adequate fixing. Always cut back to the centre of the rafter and nail the end. Never allow battens to be unsupported. Ensure replacement battens are fully graded and comply with the requirements of BS 5534.

Plain tiles

A damaged tile can be removed by raising up the neighbouring tiles with a timber wedge and sliding the tile out with the nibs clearing the top of the batten. Any nails should be removed and disposed of safely.

The replacement tile can be inserted back into position using the same technique in reverse. A dab of mastic can be placed on the underside to prevent movement. Some manufacturers may also provide proprietary fixings for replacement tiles.

Interlocking tiles

A damaged tile can be removed by first easing it up slightly so that it can be slid out with the nibs clearing the top of the batten. Timber wedges and a flat trowel will facilitate this procedure. If the damaged tile is nailed, then the neighbouring tiles should be lifted to expose the nail(s), which should be extracted carefully and disposed of safely. The replacement tile can be inserted using the same procedure in reverse.

Isolated replacement tiles should be refixed using proprietary mechanical fixings as recommended by the manufacturer or affixed to adjacent tiles that are mechanically fixed with either nails or clips, by the use of an appropriate adhesive as recommended by the manufacturer that is applied to the interlock/overlock and head-lap area.

Care should be taken to ensure that the anti-capillary bars are not bridged, and interlocks are kept clear to allow water drainage. A less aesthetic solution is to drill the left-hand bottom corner of the replacement tile in a position which aligns with the nail hole of the tile below. A suitable stainless steel drive screw with sealing washer can be used to secure the tail of the tile to the batten.

If all the damaged tiles are clipped, it may be necessary to strip back the roof to the nearest verge or valley/hip in order to re-clip the replacement tiles.

Fittings

Damaged or displaced ridge and hip tiles should be replaced individually and re-bedded and fixed with fresh mortar and mechanically fixed. Ensure the correct mix is used (typically 3:1 sand/cement) complying with BS 5534 and that all fittings are pre-wetted prior to laying.

Mechanically fix all ridge/hip tiles using the manufacturer's recommended fixings. If existing ridge and hip tiles can be satisfactorily reused, ensure that all tile surfaces are clean and old mortar is fully removed. If this cannot be achieved, then new matching replacement ridge or hip tiles should be obtained.

Valley tile replacement may necessitate stripping out adjacent tiles in order to replace existing valley tiles or trough valley units. Ensure any replacement tiles adjacent to the valley are refixed by clips and/or nails or re-bedded in mortar in accordance with BS 8000–6.

Ridge, hip, valley and verge components can also be refixed using 'dry fix' systems, which avoid the use of mortar and provide mechanical fixing. Details of proprietary dry fix systems are available from the tile manufacturer.

Roof-tile security

Tile 'chatter' in high winds is sometimes an unavoidable phenomenon that can affect all types of roofing tiles. The subsequent noise created by the tails of the tiles or slates being lifted and then dropped by the wind forces and transmitted through the roof structure is more prominent in roof designs where there is living accommodation in the roof space.

Single-lap tiles that are only nailed are more prone to 'chatter' and the additional use of tile clips may help to restrain the tail of the tile. Double-lapped tiles and slates are less prone to 'chatter' in high winds, although movement can still sometimes occur, particularly if the nails have not been driven home sufficiently.

Very often the problem is restricted to a small area of the roof where there are natural or artificial features nearby, or where a roof feature such as a chimney or dormer window might affect wind speed or create turbulence causing uplift in a particular roof area. If such a localised area can be identified, then it may be possible to address the problem by securing the tails of the tiles just in these areas using clips (for single-lap tiles).

The use of a suitable epoxy resin adhesive can also be considered, although care should be taken to ensure that the surfaces to be bonded are suitably prepared to ensure a firm surface adhesion. Adhesive should not block interlocks and anti-capillary channels, and thereby compromise the tiles' ability to shed water.

In extreme circumstances, it may be necessary to drill through the tail of the tile and then fix it into the batten of the course below using stainless steel screws or ring shank nails and sealing washers. If this course of action is proposed, then specific advice should be obtained from the manufacturer.

Mosses and lichens on tiles

The principal cause of the growth of mosses and lichens on tiled roofs is due to their rough surface that filters dirt out of rainwater. Decaying matter in the form of dead leaves or bird droppings that are deposited onto the roof also tend to lodge on the surface. Spores and seeds of mosses and lichens are blown onto the roof, or get carried there by the feet of birds, and sooner or later take root in the dirt and begin to grow.

Inevitably, the surface of some concrete tiles that have a rougher sanded or granule facing are the first to attract moss growth. Moss tends to flourish on roofs where trees are nearby and on north-facing shady slopes that remain damp longer. Steep, pitched roofs are less likely to support moss and lichen growth as they shed water more quickly than low-pitched roofs.

Moss on the surface of a roof tile will retain water longer, and where this affects the drainage of water down valleys, abutment gutters and the interlocking drainage channels of the roof tile, they should be carefully removed.

In normal circumstances, light moss growths do not damage roof products and are sometimes viewed as providing a more weathered and pleasing appearance to the roof, but where this is not the case, particularly with heavy moss growth, there are several methods that can be used to safely remove them.

Physical removal

Attempts to remove moss or lichen growths by severe scraping off the tile is not recommended as it can result in broken or damaged tiles and unsightly scrape marks on the surface of the tiles.

Careful use of a wooden scraper, however, can be effective in removing thick growths of moss. Inevitably, the process will have to be repeated in the future as the mosses and lichens return.

Removal with a proprietary toxic wash

This is perhaps the least expensive option, but very great care has to be taken with the application and is best suited to experienced and qualified operatives. Any product that is toxic to moss can also be dangerous to humans, animals and garden plants in the vicinity of the roof and its surroundings. Any residue from the use of a toxic wash applied to the surface of the tiles should be collected in a sealable container and labelled with a health and safety warning before delivery to an approved waste disposal site.

Under no circumstances should any toxic wash be allowed to drain into the groundwater courses or building drainage system (see health and safety and environmental regulations).

Toxic washes take a few days to be fully effective and should preferably be applied during a spell of dry weather, since rain may wash them off before they have had time to act. The action is hastened if thick growths are removed first, and the wash is well brushed in. Normally, one treatment is sufficient to kill the growths, but in severe cases, repeated treatment may be necessary. Even when successful, they are only likely to be effective against further moss or lichen growth for approximately two to three years.

Removal by spraying with water

If removal of the dead growths of moss or lichen is required, this can be achieved by a low-pressure (4–5 bar) jet of water, taking care not to spray against the tile laps.

On no account should a high-pressure water jet be used to clean off moss and lichen growths from concrete tiles. This will result in erosion of the tile surface, thereby reducing the potential lifespan of the roof tile.

Preventing growth using copper wire

A more permanent solution to the problem of maintaining a roof clear from moss and lichen growth is by trailing copper wires across the roof surface. These can be fixed at intervals up the roof slope, directly below the front edge of the tiles, so that with every shower of rain, the copper slowly oxidises in the atmosphere and provides the roof with a mild wash of copper sulphate, which prevents moss and lichen growth.

8.3 Major and heritage repairs

Full or part retiling

Where the condition of the roof of a historic building is poor enough to warrant stripping and retiling, it is acknowledged that English Heritage and local authority conservation officers like to see sound tiles salvaged and reused on the same roof, with any deficiencies made up with new tiles which match the existing. Manufacturers of clay and concrete products are able to assist in the provision of new tiles to match historic patterns and thereby ensure the success of such projects.

When re-covering, it is advisable to photograph the roof prior to stripping, to ensure that the existing details are properly followed. Stripping should be carried out carefully to ensure that any sound existing tiles remain undamaged so that they can be sorted according to type, size and thickness and stored for reuse. When assessing existing tiles for reuse, their likely further life should be carefully considered.

Retiling should be carried out by suitably qualified and experienced roofers, using sound tiles salvaged from the roof, with any deficiencies made up with suitable replacement tiles, matching the existing ones in type, size, thickness, colour and texture. The selection of existing tiles for reuse should be carried out with great care to try to ensure that they will have a significant future lifespan in relation to that of the new material. Manufacturers recommend that new and old tiles are not mixed, rather they be used separately on different roof sections.

In cases where the direct equivalent of the original tiles is no longer available, some manufacturers can offer bespoke tiles made to order if the quantity is sufficient for production.

Bibliography

Regulations

Approved Document B. *Fire Safety: (volume 1 – Dwellings, volume 2 – Buildings other than dwellings) - England building regulations*.

Approved Document C. *Site preparation and resistance to contaminants and moisture – England Building Regulations*.

Approved Document F. *Ventilation: Building regulation in England for the ventilation requirements to maintain indoor air quality (volume 1 – dwellings, volume 2 – Buildings other than dwellings) - England building regulations*.

Approved Document L. *Conservation of fuel and power: Building regulation in England for the ventilation requirements to maintain indoor air (volume 1 – dwellings, volume 2 – Buildings other than dwellings) - England building regulations*.

Standards

BS 516. *Prefabricated accessories for roofing – Installations for roof access – Walkways, treads and steps*.

BS 1200. *Sands for mortar for plain and reinforced brickwork, block walling and masonry*. (Superseded by BS EN 13139:2002 and declared obsolescent but remains current).

BS 1202. *Specification for nails*. (Note: Part 1 steel, Part 2 copper, Part 3 aluminium).

BS 4978. *Visual strength grading of softwood: Specification*.

BS 5250. *Management of moisture in buildings – Code of practice*.

BS 5534. *Slating and tiling for pitched roofs and vertical cladding – Code of practice*.

BS 6100–6. *Building and civil engineering – Vocabulary – Construction parts*.

BS 7883. *Personal* fall *protection equipment: Anchor systems: System design, installation and inspection: Code of practice*.

BS 7981. *Code of practice for the installation, maintenance, thorough examination and safe use of Mast Climbing Work Platforms (MCWPs)*.

BS 8000–3. *Workmanship on construction sites – Masonry – Code of practice*.

BS 8000–6. *Workmanship on construction sites – Slating and tiling of roofs and walls – Code of practice*.

BS 8104. *Code of practice for assessing exposure of walls to wind-driven rain*.

BS 8417. *Preservation of wood: Code of practice*.

BS 8612. *Dry fixed ridge, hip, and verge systems for slating and tiling – Specification*.

BS 8747. *Reinforced Bitumen Membranes (RBMs) for roofing – Guide to selection and specification*.

BS 9250. *Code of practice for design of the airtightness of ceilings in pitched roofs*.

BS EN 49. *Concrete roofing tiles and fittings for roof covering and wall cladding – Test methods*.

BS EN 131. *Ladders – Parts 1–6*.

BS EN 197–1. *Cement – Composition, specifications and conformity criteria for common cements*.

BS EN 490. *Concrete roofing tiles and fittings for roof covering and wall cladding – Product specifications*.

BS EN 517. *Prefabricated accessories for roofing: Roof safety hooks*.

BS EN 934–3. *Admixtures for concrete, mortar and grout: Admixtures for masonry mortar: Definitions, requirements, conformity and marking and labelling*.

BS EN 1004. *Mobile access and working towers made of prefabricated elements materials, dimensions, design loads, safety and performance requirements*.

BS EN 1008. *Mixing water for concrete: Specification for sampling, testing and assessing the suitability of water, including water recovered from processes in the concrete industry, as mixing water for concrete*.

BS EN 1304. *Clay roofing tiles and fittings – Product definitions and specifications*.
BS EN 1991–1–4. *Eurocode 1: Actions on structures – General actions – Wind actions*.
BS EN 1995–1–1. *Eurocode 5: Design of timber structures – General – Common rules and rules for buildings*.
BS EN 12056–3. *Gravity drainage systems inside buildings: Roof drainage, layout and calculation (AMD 17041)*.
BS EN 12810–1. *Facade scaffolds made of prefabricated components product specifications*.
BS EN 12811–1. *Temporary works equipment scaffolds: Performance requirements and general design*.
BS EN 12878. *Pigments for the colouring of building materials based on cement and/or lime: Specifications and methods of test*.
BS EN 13139. *Aggregates for mortar*.
BS EN 13141–1. *Ventilation for buildings – Performance testing of components/products for residential ventilation. Part 1: Externally and internally mounted air transfer devices*.
BS EN 13501–1. *Fire classification of construction products and building elements: Classification using data from reaction to fire tests*.
BS EN 13859–1. *Flexible sheets for waterproofing – Definitions and characteristics of underlays – Part 1: Underlays for discontinuous roofing*.
BS EN 13859–2. *Flexible sheets for waterproofing – Definitions and characteristics of underlays – Part 2: Underlays for walls*.
BS EN 60529. *Degrees of protection provided by enclosures (IP Code)*.
PD CEN/TR 15601. *Hygrothermal performance of buildings – Resistance to eind-friven rain of roof coverings with discontinuously laid small elements – Test methods*.

Other publications

ADVISORY COMMITTEE FOR ROOFSAFETY. *Safe Handling of Solar Collectors and other Large Items on roofs: Information sheet no 2: Shropshire. ACR*.
BUILDING RESEARCH ESTABLISHMENT. *Control of lichens, moulds and similar growths*. BRE Digest 370. Garston: BRE.
BUILDING RESEARCH ESTABLISHMENT. *Slate and tile roofs: Avoiding damage from aircraft wake vortices*. BRE Digest 467. Garston: BRE.
BUILDING RESEARCH ESTABLISHMENT. *Thermal insulation; avoiding risks*. BRE Report 262. Garston: BRE.
CONSTRUCTION PROJECT INFORMATION COMMITTEE (CPIC). *Common Arrangement of Work Sections (CAWS) for building works*. Alton: CPIC.
ENVIRONMENT AGENCY GB. *WM3: Waste classification – Guidance on the classification and assessment of waste*. Bristol. https://www.gov.uk/government/publications/waste-classification-technical-guidance
EUROPEAN COMMUNITIES. *The construction products regulation (EU) No 305/2011 (CPR)*. Luxembourg: Office for Official Publications of the European Communities.
GREAT BRITAIN. *The construction (design and management) regulations*. London. https://www.hse.gov.uk/construction/cdm/2015
GREAT BRITAIN. *The control of lead at work (third edition)*. London. https://www.hse.gov.uk/pubns/
GREAT BRITAIN. *The control of noise at work regulations 2005*. London. https://www.hse.gov.uk/noise/
GREAT BRITAIN. *Control of Substances Hazardous to Health (COSHH) regulations*. London. https://www.hse.gov.uk/coshh
GREAT BRITAIN. *The electricity at work regulations 1989*. HSR25. London. https://www.hse.gov.uk/pubns/books/hsr25.htm
GREAT BRITAIN. *The health and safety at work etc.act 1974*. (latest current revision). London. https://www.hse.gov.uk/legislation/hswa.htm
GREAT BRITAIN. *Health and safety in roofwork (fifth edition)*. HSG 33. London. https://www.hse.gov.uk/pubns/
GREAT BRITAIN. *The health and safety (safety signs and signals) regulations 1996*. L64 (Third Edition). London. https://www.hse.gov.uk/pubns/books/l64.htm.
GREAT BRITAIN. *The Lifting Operations and Lifting Equipment Regulations (LOLER) 1998*. London. https://books.hse.gov.uk/LOLER

GREAT BRITAIN. *The management of health and safety at work regulations 1999*. London. https://www.hse.gov.uk/pubns/hsc13.pdf
GREAT BRITAIN. *The manual handling operations regulations 1992*. L23 (Fourth Edition). London. https://www.hse.gov.uk/pubns/books/l23.htm
GREAT BRITAIN. *The personal protective equipment at work (Amendment) regulations 2022*. PPER2022. London. https://www.hse.gov.uk/ppe/ppe-regulations-2022.htm
GREAT BRITAIN. *The provision and use of work equipment regulations 1998 (PUWER)*. London. https://www.hse.gov.uk/work-equipment-machinery/puwer.htm
GREAT BRITAIN. *The work at height regulations*. London. https://www.hse.gov.uk/work-at-height/
HEALTH AND SAFETY EXECUTIVE. *Health and safety in roof work*. HSG 33. London: HSE Books.
HEALTH AND SAFETY EXECUTIVE. *Industry guidance – Working at height*. INDG401. London: HSE Books.
LEAD SHEET TRAINING ACADEMY. *Rolled lead sheet – The complete manual*. Tonbridge: LSTA.
NATIONAL FEDERATION OF ROOFING CONTRACTORS. *GN02, trafficking single-lap tiled roofs*. London: NFRC.
NATIONAL FEDERATION OF ROOFING CONTRACTORS. *GN08, underlay labelling and suitability*. London: NFRC.
NATIONAL FEDERATION OF ROOFING CONTRACTORS. *GN37, lightning protection for tiled and slated roofs*. London: NFRC.
NATIONAL FEDERATION OF ROOFING CONTRACTORS. *HSGS04, fall protection and prevention for working on pitched roofs*. London: NFRC.
NATIONAL FEDERATION OF ROOFING CONTRACTORS. *HSGS05, controlling silica when disc cutting roof tiles*. London: NFRC.
NATIONAL FEDERATION OF ROOFING CONTRACTORS. *NFRC guidance, installing pitched roofs in accordance with BS 5534*. London: NFRC.
NATIONAL FEDERATION OF ROOFING CONTRACTORS. *TB03, hooks for slating*. London: NFRC.
NATIONAL FEDERATION OF ROOFING CONTRACTORS. *TB06, pitched roof underlay*. London: NFRC.
NATIONAL FEDERATION OF ROOFING CONTRACTORS. *TB08, pitched roofing valleys – Design considerations*. London: NFRC.
NATIONAL FEDERATION OF ROOFING CONTRACTORS. *TB27, standard roofing mortar*. London: NFRC.
NATIONAL FEDERATION OF ROOFING CONTRACTORS. *TB28, inclined preformed GRP valley troughs*. London: NFRC.
NATIONAL FEDERATION OF ROOFING CONTRACTORS. *TB33, graded battens for slating and tiling*. London: NFRC.
NATIONAL FEDERATION OF ROOFING CONTRACTORS. *TB38, mortar in tiled valleys*. London: NFRC.
NATIONAL FEDERATION OF ROOFING CONTRACTORS. *TB41, solar installations on roofs*. London: NFRC.
NATIONAL FEDERATION OF ROOFING CONTRACTORS. *TB42, lime mortar for roofwork*. London: NFRC.
NATIONAL FEDERATION OF ROOFING CONTRACTORS. *TS02, durability of roofing mortar*. London: NFRC.
THE ROOF TILE ASSOCIATION. *Correct installation and safe use of slating and tiling battens,vertical tiling guide*. Stoke-on-Trent: RTA.
ROOF TILE ASSOCIATION. *A guide to the care, maintenance and repair of clay and concrete tiles used for pitched roofs and walls*. Stoke-on Trent: RTA.

Further reading

ENGLISH HERITAGE. *Practical building conservation – Roofing*. Ashgate 2013. London
HISTORIC ENGLAND. *Stone slate roofing: Technical advice note, 2005*.
KEVIN TAYLOR. *Roof tiling and slating*. Wiltshire: The Crowood Press.
NATIONAL HOUSE BUILDING COUNCIL (NHBC). *Technical standards 2024 – Chapter 7.2 pitched roofs*. NHBC. Milton Keynes. https://www.nhbc.co.uk

Index

Note: Page numbers in *italics* indicate figures, and page numbers in **bold** indicate tables in the text

For Product Safety Concerns and Information please contact our EU representative GPSR@taylorandfrancis.com Taylor & Francis Verlag GmbH, Kaufingerstraße 24, 80331 München, Germany

Batch number: 10733637

Printed by Printforce, the Netherlands